Success Series™

ALGEBRA
for
Olympiads

Problems and Solutions

J. V. R.
B.Tech., M.E., Ph.D.

ALGEBRA for Olympiads
Problems and Solutions
by J. V. R.

Copy Right © 2012 by Author

First Edition : **2012.**

Layout & D.T.P. by

Price :

Preface

With my humble presentation of this book to the readers a sincere attempt has been made to discuss elaborately on mathematical problems useful for IIT-Foundation, Regional Olympiad & Mathematical talent tests for high school students.

This book is the Vol.I of the mathematical problems books series. The Vol. II covers Number Theory, Vol. III covers Combinatorics and Vol. IV covers Plane Geometry.

I am confident that the candidates who could solve the given problems unaided would go easy with the problems set in Maths Olympiad and all competitive tests.

Each chapter consists of Synopsis, Exercise-1 and Exercise - 2. Exercise - 1 is completely solved. Students are advised to attempt sincerely twice without the help of solutions. Then they can go through the solutions. Exercise - 2 can be solved in exmination conditions.

Suggestions and comments are welcome.

J. V. R.

RMO ALGEBRA SYLLABUS

Quadratic Equations and Expressions

Systems of Linear Equations

Factorisation of Polynomials

Inequalities

Finite Series

About Mathematical Olympiads in India

Mathematics

The Mathematics Olympiad Programme in India leading to participation in the International Mathematics Olympiad is organized by the Homi Bhabha Centre for Science Education (HBCSE) on behalf of the National Board of Higher Mathematics (NBHM) of the Department of Atomic Energy (DAE).

Stages

Stage 1 : Regional Mathematical Olympiad (RMO)

For the purpose of RMO, 23 regions over the whole country have been identified, and each assigned a Regional Coordinator. Additionally, three groups (CBSE, NVS and KVS) also have a 'Regional Coordinator' each. All school students of Class XI and XII are eligible to appear for RMO. The Regional Coordinators may on their discretion admit exceptionally brilliant students of lower classes also. RMO is a 3-hour written test containing about 6 to 7 problems. On the basis of RMO, a certain number from each region is selected to appear for the Stage 2 examination. (The selected students from each region may include a maximum of 6 students of Class XII.) Regional Co-ordinators may charge some nominal fees to meet the expenses for organising the contests.

Stage 2: Indian National Mathematical Olympiad (INMO)

The INMO will be held on Sunday, January 17, 2010 between 1.00 pm and 5.00 pm. at centres in different regions. Only students selected on the basis of RMO from different regions are eligible to appear for INMO. INMO is a 4-hour written test. The question paper is set centrally and is common throughout the country. On the basis of INMO, the top 30-35 students from all over the country become INMO awardees and receives a Certificate of Merit. (A maximum 6 students of Class XII may be INMO awardees).

Stage 3: International Mathematical Olympiad Training Camp (IMOTC)

The INMO awardees are invited to a month long Training Camp held in April-May each year at the Homi Bhabha Centre for Science Education (HBCSE), Mumbai. INMO awardees of the previous year who have satisfactorily gone through postal tuition throughout the year are invited again to a second round of training (Senior Batch). The senior batch participants who successfully complete the Camp receive a prize of Rs.5,000/- in the form of books and cash. On the basis of a number of selection tests through the Camp, a team of the best six students is selected from the combined pool of junior and senior batch participants.

Stage 4: Pre-departure Training Camp for IMO

The selected team of six students goes through another round of training and orientation for about 10 days prior to the departure for IMO.

Stage 5: International Mathematical Olympiad (IMO)

The six-member team selected at the end of the Camp accompanied by a leader and a deputy leader represent the country at the IMO, held in July each year in a different member country of IMO. IMO consists of two 4 and 1/2-hour written tests held on two days. Travel to IMO venue and return takes about 2 weeks. India has been participating in IMO since 1989. Students of the Indian team who receive gold, silver and bronze medals at the IMO receive a cash prize of Rs.5000/-, Rs.4000/- and Rs.3000/- respectively during the following year at a formal ceremony at the end of the Training Camp.

Ministry of Human Resource Development (MHRD) finances international travel of the 8-member Indian delegation, while NBHM (DAE) finances the entire in-country programme and other expenditure connected with international participation.

Students aiming to go through the Mathematical Olympiad Programme leading to international participation (for the best 6) should note that RMO is the first essential step for the programme. To appear for RMO, students should get in touch with the RMO Co-ordinator of their region well in advance, for enrollment and payment of any fees, which is nominal.

Syllabus

The syllabus for mathematics olympiad (regional, national and international) is predegree college mathematics. The areas covered are arithmetic of integers, geometry, quadratic equations and expressions, trigonometry, co-ordinate geometry, systems of linear equations, permutations and combinations, factorisation of polynomials, inequalities, elementary combinatorics, probability theory and number theory, finite series and complex numbers and elementary graph theory. The syllabus does not include Calculus and Statistics. The major areas from which problems are given are number theory, geometry, algebra and combinatorics. The syllabus is in a sense spread over Class IX to Class XII levels, but the problems under each topic are of exceptionally high level in difficulty and sophistication. The difficulty level increases from RMO to INMO to IMO.

Contents

BLANK PAGE LEFT INTENTIONALLY

1. ALGEBRAIC IDENTITY

Identity: A statement of the equality of two algebraic expressions is called identity, if the equality is true for all real values of the variables involved.

1. Important Identities:

(i) $(a+b)^2 = a^2 + b^2 + 2ab$

(ii) $(a-b)^2 = a^2 + b^2 - 2ab$

(iii) $(a+b)^2 + (a-b)^2 = 2(a^2 + b^2)$

(iv) $(a+b)^2 - (a-b)^2 = 4ab$

(v) $(a+b)^2 = (a-b)^2 + 4ab$

(vi) $(a-b)^2 = (a+b)^2 - 4ab$

(vii) $(a+b+c)^2 = a^2 + b^2 + c^2 + 2(ab + bc + ca)$

(viii) $\left(\sum_{i=1}^{n} x_i\right)^2 = (x_1 + x_2 + x_3 + ... + x_n)^2$

$$= \sum_{i=1}^{n} x_i^2 + 2\Sigma x_i x_j, \ 1 \le i \le j \le n$$

(ix) $a^2 + b^2 + c^2 - ab - bc - ca = \dfrac{1}{2}\left[(a-b)^2 + (b-c)^2 + (c-a)^2\right]$

(x) $(a+b)^3 = a^3 + b^3 + 3ab(a+b)$

(xi) $(a-b)^3 = a^3 - b^3 - 3ab(a-b)$

(xii) $(a^3 + b^3) = (a+b)^3 - 3ab(a+b)$

(xiii) $a^2 + b^2 = (a+b)^2 - 2ab$

(xiv) $|a-b| = \sqrt{(a+b)^2 - 4ab}$

Problems in Algebra
2. Factorisation.

(i) $a^2 - b^2 = (a+b)(a-b)$

(ii) $a^3 - b^3 = (a-b)(a^2 + ab + b^2)$

(iii) $a^3 + b^3 = (a+b)a^2 - ab + b^2)$

(iv) $x^4 + x^2 + 1 = (x^2 + x + 1)(x^2 - x + 1)$

(v) $a^3 + b^3 + c^3 - 3abc = (a+b+c)[a^2 + b^2 + c^2 - ab - bc - ca]$

$$= \frac{1}{2}(a+b+c)[(a-b)^2 + (b-c)^2 + (c-a)^2]$$

(vi) $x^2 - (a+b)x + ab = (x-a)(x-b)$

(vii) $x^2 + (a+b)x + ab = (x+a)(x+b)$

(viii) $x^n - 1 = (x-1)(x^{n-1} + x^{n-2} + \ldots\ldots + x^2 + x + 1)$

3. Series of Natural Numbers

$$1 + 2 + 3 + \ldots\ldots + n = \frac{n(n+1)}{2}$$

$$1^2 + 2^2 + 3^2 + \ldots\ldots + n^2 = \frac{n(n+1)(2n+1)}{6}$$

$$1^3 + 2^3 + 3^3 + \ldots\ldots + n^3 = \left[\frac{n(n+1)}{2}\right]^2$$

4. Some More Identities

(i) $(a+b)(b+c)(c+a)$

$$= a^2(b+c) + b^2(c+a) + c^2(a+b) + 2abc$$

$$= a(b^2 + c^2) + b(c^2 + a^2) + c(a^2 + b^2) + 2abc$$

$$= bc(b+c) + ca(c+a) + ab(a+b) + 2abc$$

$$= (a+b+c)(ab+bc+ca) - abc$$

(ii) $(a+b+c)^3 = a^3 + b^3 + c^3 + 3(a+b)(b+c)(c+a)$

(iii) $(a+b+c+d)^2 = a^2 + b^2 + c^2 + d^2 + 2a(b+c+d) + 2b(c+d) + 2cd$

(iv) $(a+b+c+d+e)^2 = a^2 + b^2 + c^2 + d^2 + e^2$

$+2a(b+c+d+e)+2b(c+d+e)$

$+2c(d+e)+2de$

(v) $(a+b+c)^4 + a^4 + b^4 + c^4 = (a+b)^4 + (b+c)^4$

$= (c+a)^4 + 12abc(a+b+c)$

5. Conditional Equalities

If $a+b+c=0$, then

(i) $a^3 + b^3 + c^3 = 3abc$

(ii) $a^2 + b^2 + c^2 = -2(bc+ca+ab)$

(iii) $a^4 + b^4 + c^4 = 2(b^2c^2 + c^2a^2 + a^2b^2)$

$= \dfrac{1}{2}(a^2 + b^2 + c^2)^2$

(iv) $a^5 + b^5 + c^5 = -5abc(ab+bc+ca)$

$= \dfrac{5}{2}abc(a^2 + b^2 + c^2)$

$= \dfrac{5}{6}(a^2 + b^2 + c^2)(a^3 + b^3 + c^3)$

(v) $a^7 + b^7 + c^7 = 7abc(ab+bc+ca)^2$

$= \dfrac{7}{2}(a^2 + b^2 + c^2)^2(a^3 + b^3 + c^3)$

EXERCISE - 1

1. Using the identity $(a+b)^2 = a^2 + b^2 + 2ab$, find the values of (i) $(a+b+c)^2$ and (ii) $(a+b+c+d)^2$.

2. How many terms are there in $(a+b+c)^2$?

3. Find the number of terms in the expansion $(x_1 + x_2 + x_3 + \ldots \ldots x_n)^2$

Problems in Algebra

4. Prove that $1+2+3+\ldots\ldots+n=\dfrac{n(n+1)}{2}$

5. Prove the identity

$$a^2+b^2+c^2-ab-bc-ca=\frac{1}{2}\left[(a-b)^2+(b-c)^2+(c-a)^2\right]$$

6. Prove the identity

$$a^3+b^3+c^3-3abc=(a+b+c)(a^2+b^2+c^2-ab-bc-ca)$$

or factorise $a^3+b^3+c^3-3abc$

7. Prove that if $a+b+c=0$, then $a^3+b^3+c^3=3abc$

8. Factor the following:

$$(a+b+c)^3-a^3-b^3-c^3$$

9. Prove the Identity:

$$(a+b+c)^4+a^4+b^4+c^4=(a+b)^4+(b+c)^4+(c+a)^4$$

$$+12abc(a+b+c)$$

10. If $a+b+c=0$, then prove that

$$a^4+b^4+c^4=2(b^2c^2+c^2a^2+a^2b^2)=\frac{1}{2}(a^2+b^2+c^2)^2$$

11. If $a+b+c=0$, then prove that

$$a^5+b^5+c^5=-5abc(bc+ca+ab)$$

$$=\frac{5}{2}abc(a^2+b^2+c^2)$$

$$=\frac{5}{6}(a^2+b^2+c^2)(a^3+b^3+c^3)$$

12. If $a+b+c=0$, then prove that

$$a^7+b^7+c^7=7abc(ab+bc+ca)^2$$

$$=\frac{7}{12}(a^2+b^2+c^2)^2(a^3+b^3+c^3)$$

4

$$= \frac{7}{6}(a^4 + b^4 + c^4)(a^3 + b^3 + c^3)$$

13. Prove the identity,

$$(a+b)(b+c)(c+a)$$

$$= a^2(b+c) + b^2(c+a) + c^2(a+b) + 2abc$$

$$= a(b^2 + c^2) + b(c^2 + a^2) + c(a^2 + b^2) + 2abc$$

$$= bc(b+c) + ca(c+a) + ab(a+b) + 2abc$$

$$= (a+b+c)(ab+bc+ca) - abc$$

14. Prove the identity

$$(a+b+c)^3 = a^3 + b^3 + c^3 + 3(a+b)(b+c)(c+a)$$

15. Prove the identity

$$(a+b)^3 + (b+c)^3 + (c+a)^2 - 3(a+b)(b+c)(c+a)$$

$$= 2(a^3 + b^3 + c^3 - 3abc)$$

16. Prove the identity:

$$(a-b)^3 + (b-c)^3 + (c-a)^3 = 3(a-b)(b-c)(c-a)$$

17. If $b = \dfrac{c+a}{2}$, then prove $\dfrac{a}{a-b} + \dfrac{c}{c-b} = 2$

18. Show that $(a+b+c)^3 = (a+b-c)^3 + (b+c-a)^3$

$$+(c+a-b)^3 + 24abc$$

19. Find the value of $(a+b+c)^3 - a^3 - b^3 - c^3$,

if $a+b = 5, b+c = 10$ and $c+a = 15$

20. If $a = 2009, b = 2010, c = \dfrac{1}{2010}$, then find the value of

$$(a+b+c)^3 - (a+b-c)^3 - (b+c-a)^3 - (c+a-b)^3 - 23abc$$

21. If $a+b+c = 0$, then find the value of

$$\left(\frac{b-c}{a}+\frac{c-a}{b}+\frac{a-b}{c}\right)\left(\frac{a}{b-c}+\frac{b}{c-a}+\frac{c}{a-b}\right)$$

22. If $\dfrac{1}{a}+\dfrac{1}{b}+\dfrac{1}{c}=\dfrac{1}{a+b+c}$, then prove that

$$\left(\frac{1}{a^{2n+1}}+\frac{1}{b^{2n+1}}+\frac{1}{c^{2n+1}}\right)=\frac{1}{(a+b+c)^{2n+1}}$$

where n is a natural number

23. Show that

$$\left(x+\frac{1}{x}\right)\left(x-\frac{1}{x}\right)\left(x^2+\frac{1}{x^2}+1\right)\left(x^2+\frac{1}{x^2}-1\right)=x^6-\frac{1}{x^6}$$

(Croatian mathematical olympiad 2002)

24. Show that

$$(a^2-bc)^3+(b^2-ca)^3+(c^2-ab)^3$$

$$-3(a^2-bc)(b^2-ca)(c^2-ab)$$

$$=(a^3+b^3+c^3-3abc)^2$$

25. If x,y,z are distinct real numbers such that

$$\frac{x}{y-z}+\frac{y}{z-x}+\frac{z}{x-y}=0 ,\text{ then prove}$$

$$\frac{x}{(y-z)^2}+\frac{y}{(z-x)^2}+\frac{z}{(x-y)^2}=0$$

26. Prove that

$$\frac{(y-z)^5+(z-x)^5+(x-y)^5}{5}$$

$$=\frac{(y-z)^3+(z-x)^3+(x-y)^3}{3}\times\frac{(y-z)^2+(z-x)^2+(x-y)^2}{2}$$

27. Prove that $\dfrac{a}{1+a}+\dfrac{b}{1+b}+\dfrac{c}{1+c}=1$,

if $x = by + cz,\ y = cz + ax$ and $z = ax + by$.

EXERCISE - 1
Solutions

1. (i) $(a+b+c)^2 = [(a+b)+c]^2$

   $= (a+b)^2 + c^2 + 2(a+b)c$

   $= a^2 + b^2 + 2ab + c^2 + 2ac + 2bc$

   $= a^2 + b^2 + c^2 + 2ab + 2bc + 2ca$

   (ii) $(a+b+c+d)^2 = [(a+b)+(c+d)]^2$

   $= (a+b)^2 + (c+d)^2 + 2(a+b)(c+d)$

   $= a^2 + b^2 + 2ab + c^2 + d^2 + 2cd + 2(ac+ad+bc+bd)$

   $= a^2 + b^2 + c^2 + d^2 + 2ab + 2ac + 2ad + 2bc + 2cd$

2. The numbers of terms in a^2 is 1.
   The number of terms in $(a+b)^2$ is $3 = 1 + 2$
   The number of terms in $(a+b+c)^2$ is $1 + 2 + 3 = 6$
   Check this value?

3. The number of terms

   $$S = 1 + 2 + 3 + \ldots\ldots n = \frac{n(n+1)}{2}$$

4. Let $S = 1 + 2 + 3 + \ldots\ldots + 3 + 2 + 1$.

   Adding these two equations, we have

   $2s = (n+1) + (n+1) + \ldots + (n+1) + (n+1) + (n+1)$

   $\Rightarrow \qquad 2s = n(n+1)$

   $\Rightarrow \qquad s = \dfrac{n(n+1)}{2}$

5. $a^2 + b^2 + c^2 - ab - bc - ca = \dfrac{1}{2}\left[(a-b)^2 + (b-c)^2 + (c-a)^2\right]$

$$= \frac{1}{2}\left[a^2 + a^2 + b^2 + b^2 + c^2 + c^2 - 2ab - 2bc - 2ca\right]$$

$$= \frac{1}{2}\left[(a^2 + b^2 - 2ab) + (b^2 + c^2 - 2bc) + (c^2 + a^2 - 2ca)\right]$$

$$= \frac{1}{2}\left[(a-b)^2 + (b-c)^2 + (c-a)^2\right]$$

Hence Proved.

6. $a^3 + b^3 + c^3 - 3abc$

$$= a^3 + 3ab(a+b) + b^3 + c^3 - 3abc - 3ab(a+b)$$

$$= a^3 + 3a^2b + 3ab^2 + b^3 + c^3 - 3ab(c+a+b)$$

$$= (a+b)^3 + c^3 - 3ab(a+b+c)$$

$$= \left[(a+b)+c\right]\left[(a+b)^2 - (a+b)c + c^2\right] - 3ab(a+b+c)$$

$$= (a+b+c)\left[(a+b)^2 - (a+b)c + c^2 - 3ab\right]$$

$$= (a+b+c)(a^2 + b^2 + 2ab - ac - bc + c^2 - 3ab)$$

$$= (a+b+c)(a^2 + b^2 + c^2 - ab - bc - ca)$$

Alternate Method:

Let P denote the polynomial with roots a,b,c :

$$P(x) = x^3 - (a+b+c)x^2 + (ab+bc+ca)x - abc$$

The equation $P(x) = 0$ is satisfied by a,b,c then

$$a^3 - (a+b+c)a^2 + (ab+bc+ca)a - abc = 0$$

$$b^3 - (a+b+c)b^2 + (ab+bc+ca)b - abc = 0$$

$$c^3 - (a+b+c)c^2 + (ab+bc+ca)c - abc = 0$$

Adding up these three equalities, then

$$a^3 + b^3 + c^3 - (a+b+c)(a^2 + b^2 + c^2)$$

$$+(ab+bc+ca)(a+b+c) - 3abc = 0$$

$\Rightarrow \quad a^3 + b^3 + c^3 - 3abc = (a+b+c)(a^2+b^2+c^2-ab-bc-ca)$

7. $\qquad a^3 + b^3 + c^3 - 3abc$

$\qquad = (a+b+c)(a^2+b^2+c^2-ab-bc-ca) \quad (\because \text{ from problem 6})$

$\because \qquad a+b+c = 0$

$\therefore \qquad a^3 + b^3 + c^3 - 3abc = 0$

$\Rightarrow \qquad a^3 + b^3 + c^3 = 3abc$

Alternate Method:

$\because \qquad a+b+c = 0$

We have,

$0 = (a+b+c)^3$

$\qquad = a^3 + b^3 + c^3 + 3ab(a+b) + 3bc(b+c) + 3ca(c+a) + 6abc$

$\qquad = a^3 + b^3 + c^3 - 3abc - 3abc - 3abc + 6abc$

$\qquad = a^3 + b^3 + c^3 - 3abc \Rightarrow a^3 + b^3 + c^3 = 3abc \cdot$

8. $\qquad (a+b+c)^3 - a^3 - b^3 - c^3$

$\qquad = \left[(a+b+c)^3 - a^3\right] - (b^3 - c^3)$

$= \left[(a+b+c) - a\right]\left[(a+b+c)^2 + a(a+b+c) + a^2\right] - (b+c)(b^2 - bc + c^2)$

$\qquad = (b+c)\left[\left\{(a+b+c)^2 - b^2\right\} + a(a+c) + (ab+bc) + (a^2-c^2)\right]$

$\qquad = (b+c)\left[\begin{array}{l}\left\{(a+b+c)-b\right\}\left\{(a+b+c)+b\right\} + a(a+c) \\ + b(a+c) + (a+c)(a-c)\end{array}\right]$

$\qquad = (b+c)(a+c)(a+b+c+b+a+b+a-c)$

$\qquad = 3(b+c)(a+c)(a+b)$

Alternate Method:

$\qquad (a+b+c)^3 - a^3 - b^3 - c^3$

Let $\qquad a = x, \qquad$ then

9

$$(x+b+c)^3 - x^3 - b^3 - c^3$$

If $x = -b$, then the expression vanishes;

∴ The equation $(x+b+c)^3 - x^3 - b^3 - c^3 = 0$ has a root $x = -b$.

∴ $(x+b+c)^3 - x^3 - b^3 - c^3$ is divisible by $x+b$.

Putting $x = a$, then we have

$(a+b+c)^3 - a^3 - b^3 - c^3$ which is divisible by $a+b$.

Similarly, we can show that

$(a+b+c)^3 - a^3 - b^3 - c^3$ is also divisible by $a+c$ and $b+c$.

∵ the three factors are relatively prime,

we have, $(a+b+c)^3 - a^3 - b^3 - c^3 = k(a+b)(a+c)(b+c)$

Put $a = 0$, $b = c = 1$, then we get $k = 3$.

9. If $P(x)$ is a polynomial if $P(a) = 0$ then $(x-a)$ is a factor.
    Consider,

    $$f(a,b,c) = (a+b+c)^4 + a^4 + b^4 + c^4 - (a+b)^4 - (b+c)^4 - (c+a)^4$$

    Put $a = 0$, then

    $$f(0,b,c) = (b+c)^4 + b^4 + c^4 - b^4 - (b+c)^4 - c^4 = 0$$

    ∴ a is a factor.

    Similarly, b, c are also factors.

    The degree of the expression = 4

    But the degree of the factors = 3

    ∴   There will be an algebraic factor which is symmetric in a, b, c of first degree and a numerical factor.

∴ $(a+b+c)^4 + a^4 + b^4 + c^4 - (a+b)^4 - (b+c)^4 - (c+a)^4 \equiv kabc(a+b+c)$

Put $a = b = c = 1$

∴ $3^4 + 1 + 1 + 1 - 2^4 - 2^4 - 2^4 \equiv k(3)$

⇒ $k = 12$

∴ $(a+b+c)^4 + a^4 + b^4 + c^4$

$$= (a+b)^4 + (b+c)^4 + (c+a)^4 + 12abc(a+b+c)$$

10. $\because$ $(a+b+c) = 0$

$$a^2 + b^2 + c^2 = -2(bc + ca + ab)$$

Squaring both sides, we get

$$(a^2 + b^2 + c^2)^2 = 4(bc + ca + ab)^2$$

$$= 4\{b^2c^2 + c^2a^2 + a^2b^2 + 2(bc.ca + ca.ab + ab.bc)\}$$

$$= 4\{b^2c^2 + c^2a^2 + a^2b^2 + 2abc(a+b+c)\} \qquad (\because\ a+b+c=0)$$

$$= 4(b^2c^2 + c^2a^2 + a^2b^2)$$

$\therefore$ $$2(b^2c^2 + c^2a^2 + a^2b^2) = \frac{1}{2}\left(a^2 + b^2 + c^2\right)^2$$

$\because$ $$(a^2 + b^2 + c^2)^2 = 4(b^2c^2 + c^2a^2 + a^2b^2)$$

$\Rightarrow$ $$a^4 + b^4 + c^4 + 2(a^2b^2 + b^2c^2 + c^2a^2)$$

$$= 4(b^2c^2 + c^2a^2 + a^2b^2)$$

$\Rightarrow$ $$a^4 + b^4 + c^4 = 2(b^2c^2 + c^2a^2 + a^2b^2)$$

11. $\because$ $$a + b + c = 0 \Rightarrow a + b = -c$$

$\Rightarrow$ $$(a+b)^5 = -c^5$$

$\Rightarrow$ $$a^5 + 5a^4b + 10a^3b^2 + 10a^2b^3 + 5ab^4 + b^5 = -c^5$$

$$= -5ab(a^3 + 2a^2b + 2ab^2 + b^3)$$

$$= -5ab(a+b)(a^2 + ab + b^2) \qquad (\because a+b=-c)$$

$$= +5abc(a^2 + ab + b^2)$$

$$= 5abc\left[(a+b)^2 - ab\right]$$

$$= 5abc\left[(a+b)(-c) - ab\right]$$

$\Rightarrow$ $$a^5 + b^5 + c^5 = 5abc[-ac - bc - ab]$$

$$\therefore \qquad a^5 + b^5 + c^5 = -5abc(ab + bc + ca)$$

$$\because \qquad a + b + c = 0 \implies a^2 + b^2 + c^2 = -2(ab + bc + ca)$$

$$\implies \qquad -(ab + bc + ca) = \frac{a^2 + b^2 + c^2}{2}$$

$$a^5 + b^5 + c^5 = \frac{5}{2}(a^2 + b^2 + c^2)abc$$

$$a^5 + b^5 + c^5 = \frac{5}{6}(a^2 + b^2 + c^2).3abc$$

$$\therefore \qquad a + b + c = 0 \implies a^3 + b^3 + c^3 = 3abc$$

$$\therefore \qquad a^5 + b^5 + c^5 = \frac{5}{6}(a^2 + b^2 + c^2)(a^3 + b^3 + c^3)$$

12. Since $a + b + c = 0 \implies a + b = -c$

$$\therefore \qquad (a + b)^7 = (-c)^7 = -c^7$$

$$\therefore \qquad (a + b)^5 (a + b)^2 = -c^7$$

$$a^5 + 5a^4b + 10a^3b^2 + 10a^2b^3 + 5ab^4 + b^5)(a^2 + b^2 + 2ab) = -c^7$$

$$\implies a^7 + 7a^6b + 21a^5b^2 + 35a^4b^3 + 35a^3b^4 + 21a^2b^5$$

$$+ 7ab^6 + b^7 + c^7 = 0$$

$$\therefore \qquad a^7 + b^7 + c^7 = -7ab(a^5 + 3a^4b + 5a^3b^2 + 5a^2b^3 + 3ab^4 + b^5)$$

$$= -7ab(a + b)(a^4 + 2a^3b + 3a^2b^2 + 2ab^3 + b^4)$$

$$= 7abc(a^2 + ab + b^2)^2$$

$$= 7abc(ab + bc + ca)^2$$

$$\because \qquad a^7 + b^7 + c^7 = 7abc(ab + bc + ca)^2$$

$$\because \qquad a^3 + b^3 + c^3 = 3abc$$

$$\implies \qquad abc = \frac{a^3 + b^3 + c^3}{3}$$

and $\qquad bc + ca + ab = \dfrac{1}{2}(a^2 + b^2 + c^2)$

$\therefore \qquad a^7 + b^7 + c^7 = \dfrac{7}{12}(a^3 + b^3 + c^3)(a^2 + b^2 + c^2)^2$

$\therefore \qquad \dfrac{(a^2 + b^2 + c^2)^2}{2} = (a^4 + b^4 + c^4)$

$\therefore \qquad a^7 + b^7 + c^7 = \dfrac{7}{6}(a^3 + b^3 + c^3)(a^4 + b^4 + c^4)$

13. $\quad (a+b)(b+c)(c+a)$

$= (b+c)\big[(a+b)(c+a)\big] = (b+c)\big[a^2 + a(b+c) + bc\big]$

$= a^2(b+c) + a(b+c)^2 + bc(b+c)$

$= a^2(b+c) + a(b^2 + 2bc + c^2) + b^2c + bc^2$

$= a^2(b+c) + b^2(c+a) + c^2(a+b) + 2abc$

$\therefore \qquad (a+b)(b+c)(c+a)$

$= a^2(b+c) + b^2(c+a) + c^2(a+b) + 2abc \qquad \text{... (i)}$

$= \underbrace{a^2b + a^2c + b^2c + ab^2 + c^2a + bc^2 + 2abc} \qquad \text{... (a)}$

$a(b^2 + c^2) + b(c^2 + a^2) + c(a^2 + b^2) + 2abc \qquad \text{... (ii)}$

from (a); $(a+b)(b+c)(c+a)$

$= \underbrace{a^2b + a^2c + b^2c + ab^2 + c^2a + bc^2 + 2abc}$

$= ab(a+b) + ac(a+c) + bc(b+c) + 2abc \qquad \text{... (iii)}$

from (iii)

$(a+b)(b+c)(c+a)$

$= ab(a+b) + bc(b+c) + ca(c+a) + 2abc$

$= ab(a+b) + abc + bc(b+c) + abc + ca(c+a) + abc - abc$

$$= ab(a+b+c) + bc(a+b+c) + ca(a+b+c) - abc$$

$$= (a+b+c)(ab+bc+ca) - abc \qquad \text{... (iv)}$$

14. $(a+b+c)^3 = \left[(a+b)+c\right]^3$

$$= (a+b)^3 + 3(a+b)c(a+b+c) + c^3$$

$$= a^3 + b^3 + 3ab(a+b) + 3c(a+b)^2 + 3c^2(a+b) + c^3$$

$$= a^3 + b^3 + c^3 + 3(a+b)[ab + c(a+b) + c^2]$$

$$= a^3 + b^3 + c^3 + 3(a+b)[ab + bc + ca + c^2]$$

$$= a^3 + b^3 + c^3 + 3(a+b)[b(a+c) + c(a+c)]$$

$$= a^3 + b^3 + c^3 + 3(a+b)(b+c)(c+a)$$

15. From identify,

$$a^3 + b^3 + c^3 - 3abc = (a+b+c)$$

$$(a^2 + b^2 + c^2 - ab - bc - ca) \qquad \text{... (i)}$$

we have, $(a+b)^3 + (b+c)^3 + (c+a)^3 - 3(a+b)(b+c)(c+a)$

$$= (a+b+b+c+c+a)\begin{bmatrix} (a+b)^2 + (b+c)^2 + (c+a)^2 \\ -(a+b)(b+c) - (b+c)(c+a) \\ -(c+a)(a+b)] \end{bmatrix}$$

$$= 2(a+b+c)\begin{bmatrix} (a+b)\{a+b-b-c\} + (b+c)\{b+c-c-a\} \\ +(c+a)\{c+a-a-b\} \end{bmatrix}$$

$$= 2(a+b+c)\left[(a+b)(a-c) + (b+c)(b-a) + (c+a)(c-b)\right]$$

$$= 2(a+b+c)\begin{bmatrix} a^2 - ac + ab - bc + b^2 - ab + bc - ac \\ +c^2 - bc + ac - ab \end{bmatrix}$$

$$= 2(a+b+c)(a^2 + b^2 + c^2 - ab - bc - ca) \qquad [\because \text{ from (i)}]$$

$$= 2(a^3 + b^3 + c^3 - 3abc)$$

16.   Let $x = a - b$; $y = b - c$; $z = c - a$

       $\Rightarrow$ $x + y + z = 0$

       $\therefore$ $x^3 + y^3 + z^3 = 3xyz$

       Substituting the values, we get

$$(a-b)^3 + (b-c)^3 + (c-a)^3 = 3(a-b)(b-c)(c-a)$$

17.   Given $b = \dfrac{c+a}{2}$

       $\Rightarrow$ $2b = c + a \Rightarrow a - b = b - c$

       $\therefore$ Given equation $\dfrac{a}{a-b} + \dfrac{c}{c-b}$

$$= \frac{a}{b-c} - \frac{c}{b-c}$$

$$= \frac{a-c}{b-c} = \frac{2b-c-c}{b-c} \qquad \left(\because a = 2b - c \right)$$

$$= \frac{2(b-c)}{(b-c)} = 2$$

18.   Let $a + b - c = z$, $b + c - a = x$, $c + a - b = y$.

       Then we have

$$x + y + z = a + b - c + b + c - a + c + a - b$$

$$= a + b + c$$

$$x + z = a + b - c + b + c - a = 2b$$

$$x + y = b + c - a + c + a - b = 2c$$

$$y + z = c + a - b + a + b - c = 2a$$

       $\therefore$ $(a+b+c)^3 = (x+y+z)^3$

$$= x^3 + y^3 + z^3 + 3(x+y)(y+z)(z+x)$$

$$= (b+c-a)^3 + (c+a-b)^3 + (a+b-c)^3$$

$$+3(2c)(2a)(2b)$$

$$= (b+c-a)^3 + (c+a-b)^3 + (a+b-c)^3 + 24abc$$

19. From the identity,

$$(a+b+c)^3 = a^3 + b^3 + c^3 + 3(a+b)(b+c)(c+a)$$

$$\Rightarrow (a+b+c)^3 - a^3 - b^3 - c^3$$

$$= 3(a+b)(b+c)(c+a) = 3(5)(10)(15) = 2250$$

20. From the problem 19,

$$(a+b+c)^3 - (a+b-c)^3 - (b+c-a)^3 - (c+a-b)^3 - 23abc = abc$$

$$= 2009 \times 2010 \times \frac{1}{2010}$$

$$= 2009$$

21. $$\frac{b-c}{a} + \frac{c-a}{b} + \frac{a-b}{c}$$

$$= \frac{b^2 c - bc^2 + ac^2 - a^2 c + a^2 b - ab^2}{abc}$$

$$= \frac{c^2(a-b) + ab(a-b) - (ac+bc)(a-b)}{abc}$$

$$= \frac{(a-b)(c^2 + ab - ac - bc)}{abc}$$

$$= \frac{(a-b)[c(c-a) - b(c-a)]}{abc}$$

$$= -\frac{(a-b)(b-c)(c-a)}{abc} \qquad \qquad \text{... (i)}$$

$$\frac{a}{b-c} + \frac{b}{c-a} + \frac{c}{a-b} = ?$$

For this let $x = b-c$, $y = c-a$ and $z = a-b$

Then, $y - z = c - a - a + b = b + c - 2a$

$$\because \quad a+b+c=0 \Rightarrow b+c=-a$$

$$\therefore \quad y-z=-3a$$

$$\Rightarrow \quad a=-\frac{y-z}{3}$$

Similarly, $b=-\dfrac{z-x}{3}$

and $\quad c=-\dfrac{x-y}{3}$

$$\therefore \quad \frac{a}{b-c}+\frac{b}{c-a}+\frac{c}{a-b}$$

$$=-\frac{1}{3}\left(\frac{y-z}{x}+\frac{z-x}{y}+\frac{x-y}{z}\right)$$

from (i), $\quad =-\dfrac{1}{3}\left[-\dfrac{(x-y)(y-z)(z-x)}{xyz}\right]$

$$=\frac{1}{3}\left[\frac{(-3c)(-3a)(-3b)}{(b-c)(c-a)(a-b)}\right]$$

$$=-9\frac{abc}{(a-b)(b-c)(c-a)}$$

$\therefore$ If $a+b+c=0$, then

$$\left(\frac{b-c}{a}+\frac{c-a}{b}+\frac{a-b}{c}\right)\left(\frac{a}{b-c}+\frac{b}{c-a}+\frac{c}{a-b}\right)$$

$$=\left[-\frac{(a-b)(b-c)(c-a)}{abc}\right]\left[-9\frac{abc}{(a-b)(b-c)(c-a)}\right]=9$$

Alternate Method:

$$\frac{b-c}{a}+\frac{c-a}{b}+\frac{c-b}{c}=-\frac{(a-b)(b-c)(c-a)}{abc}$$

$$\frac{a}{b-c}+\frac{b}{c-a}+\frac{c}{a-b}$$

$$=\frac{a(c-a)(a-b)+b(b-c)(a-b)+c(b-c)(c-a)}{(b-c)(c-a)(a-b)}$$

$$a(c-a)(a-b)=-a^3+a^2(b+c)-abc \qquad\qquad (\because b+c=-a)$$

$$=-a^3+a^2(-a)-abc$$

$$=-2a^3-abc$$

Similarly, $\qquad\qquad b(b-c)(a-b)=-2b^3-abc$

$$c(b-c)(c-a)=-2c^3-abc$$

$$\therefore \quad \frac{a}{b-c}+\frac{b}{c-a}+\frac{c}{a-b}$$

$$=\frac{-2a^3-abc-2b^3-abc-2c^3-abc}{(a-b)(b-c)(c-a)}$$

$$=\frac{-2(a^3+b^3+c^3)-3abc}{(a-b)(b-c)(c-a)} \qquad \left(\because a^3+b^3+c^3=3abc\right)$$

$$=\frac{-2(3abc)-3abc}{(a-b)(b-c)(c-a)}=-\frac{9abc}{(a-b)(b-c)(c-a)}$$

$$\therefore \quad \left(\frac{a-b}{c}+\frac{b-c}{a}+\frac{c-a}{b}\right)\left(\frac{a}{b-c}+\frac{b}{c-a}+\frac{c}{a-b}\right)$$

$$=\left[-\frac{(a-b)(b-c)(c-a)}{abc}\right]\left[-9\frac{abc}{(a-b)(b-c)(c-a)}\right]$$

$$=9$$

22. $\quad \dfrac{1}{a}+\dfrac{1}{b}+\dfrac{1}{c}=\dfrac{1}{a+b+c}$

$$\Rightarrow \frac{bc+ca+ab}{abc}=\frac{1}{a+b+c}$$

$\Rightarrow (a+b+c)(ab+bc+ca) = abc$

from problem 13, we have

$(a+b)(b+c)(c+a) = (a+b+c)(ab+bc+ca) - abc = 0$

$\Rightarrow a+b=0 \quad or \quad b+c=0 \quad or \quad c+a=0$

If $b+c=0$, then $c=-b$,

$$\therefore \quad \left(\frac{1}{a^{2n+1}} + \frac{1}{b^{2n+1}} + \frac{1}{c^{2n+1}} \right) = \frac{1}{(a+b+c)^{2n+1}}$$

$$\Rightarrow \quad \left(\frac{1}{a^{2n+1}} + \frac{1}{b^{2n+1}} - \frac{1}{b^{2n+1}} \right) = \frac{1}{(a+0)^{2n+1}}$$

$$\Rightarrow \quad \frac{1}{a^{2n+1}} = \frac{1}{a^{2n+1}}$$

Hence the result.

24. $\text{L.H.S} = \left(x + \frac{1}{x} \right) \left(x^2 + \frac{1}{x^2} - 1 \right) \left(x - \frac{1}{x} \right) \left(x^2 + \frac{1}{x^2} + 1 \right)$

$$= \left(x^3 + \frac{1}{x^3} \right) \left(x^3 - \frac{1}{x^3} \right) \quad \begin{bmatrix} \because a^3 + b^3 = (a+b)(a^2+b^2-ab) \\ a^3 - b^3 = (a-b)(a^2+b^2+ab) \end{bmatrix}$$

$$= x^6 - \frac{1}{x^6} \quad \quad \left[\because (a+b)(a-b) = a^2 - b^2 \right]$$

$= \text{R.H.S.}$

25. Let $a^2 - bc = x, \ b^2 - ca = y, \ z = c^2 - ab$

$x + y + z = (a^2 + b^2 + c^2 - ab - bc - ca)$

$y - z = b^2 - ca - c^2 + ab$

$\qquad = b^2 - c^2 + ab - ac$

$\qquad = (b+c)(b-c) + a(b-c)$

$\qquad = (a+b+c)(b-c)$

Similarly,

$$z - x = (a + b + c)(c - a)$$

$$x - y = (a + b + c)(a - b)$$

$$\therefore \quad (y - z)^2 + (z - x)^2 + (x - y)^2$$

$$= (a + b + c)^2 \left\{ (b - c)^2 + (c - a)^2 + (a - b)^2 \right\}$$

$$\therefore \quad x^3 + y^3 + z^3 - 3xyz$$

$$= \frac{1}{2}(x + y + z)\left\{ (y - z)^2 + (z - x)^2 + (x - y)^2 \right\}$$

$$= \frac{1}{2}(a^2 + b^2 + c^2 - ab - bc - ca)$$

$$\left[(a + b + c)^2 \times \left\{ (b - c)^2 + (c - a)^2 + (a - b)^2 \right\} \right]$$

$$= (a^2 + b^2 + c^2 - ab - bc - ca)^2 (a + b + c)^2$$

$$= \left\{ (a + b + c)(a^2 + b^2 + c^2 - ab - bc - ca) \right\}^2$$

$$= (a^3 + b^3 + c^3 - 3abc)^2 = \text{R.H.S}$$

26. $\left(\dfrac{x}{y - z} + \dfrac{y}{z - x} + \dfrac{z}{x - y} \right)\left(\dfrac{1}{y - z} + \dfrac{1}{z - x} + \dfrac{1}{x - y} \right)$

$$= \frac{x}{(y - z)^2} + \frac{y}{(z - x)^2} + \frac{z}{(x - y)^2}$$

$$+ \frac{x}{(y - z)(z - x)} + \frac{y}{(z - x)(x - y)} + \frac{z}{(x - y)(y - z)}$$

$$+ \frac{x}{(y - z)(x - y)} + \frac{y}{(z - x)(y - z)} + \frac{z}{(x - y)(z - x)}$$

$$= \frac{x}{(y - z)^2} + \frac{y}{(z - x)^2} + \frac{z}{(x - y)^2}$$

$$+ \frac{x + y}{(y - z)(z - x)} + \frac{y + z}{(x - y)(z - x)} + \frac{z + x}{(x - y)(y - z)}$$

$$= \frac{x}{(y-z)^2} + \frac{y}{(z-x)^2} + \frac{z}{(x-y)^2}$$

$$+ \frac{(x+y)(x-y) + (y+z)(y-z) + (z+x)(z-x)}{(y-z)(z-x)(x-y)}$$

$$= \frac{x}{(y-z)^2} + \frac{y}{(z-x)^2} + \frac{z}{(x-y)^2} + 0$$

$$= \frac{x}{(y-z)^2} + \frac{y}{(z-x)^2} + \frac{z}{(x-y)^2}$$

Hence the result.

27. Let $s = ax + by + cz$

$\therefore \qquad x + ax = s$

$\Rightarrow \qquad x = \dfrac{s}{1+a}$

Similarly, $y = \dfrac{s}{1+b}$ and $z = \dfrac{s}{1+c}$

$\therefore \qquad s = ax + by + cz$

$$= \frac{as}{1+a} + \frac{bs}{1+b} + \frac{cs}{1+c}$$

$$= s\left(\frac{a}{1+a} + \frac{b}{1+b} + \frac{c}{1+c} \right)$$

$\Rightarrow \qquad \dfrac{a}{1+a} + \dfrac{b}{1+b} + \dfrac{c}{1+b} = 1.$

EXERCISE - 2 [Time: 3hrs]

1. A leaf is torn from a book. The sum of the pages on the remaining pages is 15000. What are the page numbers on the torn leaf?

Problems in Algebra

2. If $a+b+c=0$ prove that

$$(bc+ca+ab)^2 = b^2c^2 +c^2a^2 +a^2b^2 = \frac{1}{4}(a^2 +b^2 +c^2)^2$$

3. Show that

$$(a^2 +b^2)(x^2 +y^2) = (ax-by)^2 +(bx+ay)^2$$

4. Find the least natural number whose last digit is 7 such that it becomes 5 times larger when this last digit is carried to the beginning of this number.

5. If $x=5+2\sqrt{6}$, find the values of $\sqrt{x} +\dfrac{1}{\sqrt{x}}$ and $\sqrt{x} -\dfrac{1}{\sqrt{x}}$.

6. Prove that $(a+b+c+d)^2 +(a+b-c-d)^2 +(a+c-b-d)^2 +$
$$+(a+d-b-c)^2 =4(a^2 +b^2 +c^2 +d^2)$$

7. Prove that

$$\left[(a-b)^2 +(b-c)^2 +(c-a)^2\right]^2 =2\left[(a-b)^4 +(b-c)^4 +(c-a)^4\right]$$

8. Prove that

$$\frac{(x-y)^5 +(y-z)^5 +(z-x)^5}{5}$$

$$=(x-y)(y-z)(z-x)\frac{\left[(x-y)^2 +(y-z)^2 +(z-x)^2\right]}{2}$$

9. Prove that

$$(x-y)^3 +(y-z)^3 +(z-x)^3 =3(x-y)(y-z)(z-x)$$

EXERCISE - 2
Answers

1. 25, 26
4. 142857
5. $2\sqrt{3}, 2\sqrt{2}$

22

2. LINEAR EQUATIONS IN ONE VARIABLE

The general equation of the first degree with the unknown x has the form $ax + b = 0$

where a and b are given numbers $(a \neq 0)$.

To solve the first degree equation means to reduce it to this form, and then the expression for the root is

$$x = -\frac{b}{a}$$

EXERCISE - 1

1. Solve the equation $(x - 2009)^3 + (x - 2010)^3 + (x - 2011)^3$

    $$= 3(x - 2009)(x - 2010)(x - 2011)$$

2. Solve $\cfrac{6}{7 - \cfrac{6}{7 - \cfrac{6}{7 - x}}} = 1$

3. Solve $\dfrac{a+b-x}{c} + \dfrac{a+c-x}{b} + \dfrac{c+b-x}{a} + \dfrac{4x}{a+b+c} = 1$

4. If a, b, c are integers and $a + b\sqrt{2} + c\sqrt{3} = 0$ show that $a = b = c = 0$

5. Solve $x^{\sqrt{x}} = \sqrt{x^x}$

6. Solve $3^x + 4^x = 5^x$

7. Write down preferably by inspection, the solution of the equation

    $$\frac{x-ab}{a+b} + \frac{x-ac}{a+c} = \frac{x-bc}{b+c} = a+b+c$$

8. Find the values of x which satisfies both the equations

    $(a-b)x^2 + (b-c)x + (c-a) = 0$ and

    $(c-a)x^2 + (b-c)x + (a-b) = 0$

9. Find the smallest value of $x^2 + 8x$.

10. Solve

Problems in Algebra

$$\sqrt{x+2\sqrt{x+\ldots\ldots+2\sqrt{x+2\sqrt{x}}}} = x$$

11. Solve $\dfrac{1}{x+3} + \dfrac{1}{x+7} = \dfrac{1}{x+1} + \dfrac{1}{x+9}$

12. Solve $\dfrac{a}{x-a} + \dfrac{b}{x-b} = \dfrac{a+b}{x-a-b}$

13. Solve $\dfrac{x-a}{b+c} + \dfrac{x-b}{c+a} + \dfrac{x-c}{a+b} = 3$

14. Solve $\dfrac{x-a-b}{c} + \dfrac{x-b-c}{a} + \dfrac{x-c-a}{b} = 3$ given that $ab+bc+ca \neq 0$ and none of a,b,c is zero.

15. Solve the equation.

$$\sqrt{3x+4} + \sqrt{x-4} = 2\sqrt{x}$$

16. Solve the equation

$$\frac{9-x}{x-4} = \frac{5}{x-4} - 3$$

17. Solve $\sqrt{x+\sqrt{x+11}} + \sqrt{x-\sqrt{x+11}} = 4$

18. Solve the equation $\sqrt{17+x} - \sqrt{17-x} = 2$

19. Solve $\sqrt{x-\sqrt{x-2}} = \sqrt{x+\sqrt{x-2}} = 2$

20. Solve $(x^2-4)\sqrt{x+1} = 0$

21. Solve $\sqrt[3]{1+\sqrt{x}} + \sqrt[3]{1-\sqrt{x}} = 2$

22. Solve $\sqrt[3]{x+1} + \sqrt[3]{x+2} + \sqrt[3]{x+3} = 0$

23. Let x be a real number such that $\sqrt[3]{x} + \dfrac{1}{\sqrt[3]{x}} = 3$

Find the value of $x^3 + \dfrac{1}{x^3}$

24. Solve $\dfrac{x}{a+b} + 1 = \dfrac{x}{a-b} + \dfrac{a-b}{a+b}$

25. Solve $\dfrac{m(x+a)}{x+b} + \dfrac{n(x+b)}{x+a} = m+n$

26. Solve $\dfrac{\sqrt{x+3} - \sqrt{x+2}}{\sqrt{x+3} + \sqrt{x+2}} = \dfrac{4}{5}$

EXERCISE - 1
Solutions

1. We know, if $a+b+c = 0$, then $a^3 + b^3 + c^3 = 3abc$

   Similarly, if $a^3 + b^3 + c^3 = 3abc$ then $a+b+c = 0$

   $\therefore \qquad x - 2009 + x - 2010 + x - 2011 = 0$

   $\Rightarrow \qquad 3x - 6030 = 0$

   $\Rightarrow \qquad x = \dfrac{6030}{3} = 2010$

2. From the equation,

$$6 = 7 - \cfrac{6}{7 - \cfrac{6}{7-x}}$$

$$\Rightarrow \qquad \cfrac{6}{7 - \cfrac{6}{7-x}} = 7 - 6 = 1$$

$$\Rightarrow \qquad 6 = 7 - \cfrac{6}{7-x}$$

$$\Rightarrow \qquad \cfrac{6}{7-x} = 7 - 6 = 1$$

$$\Rightarrow \qquad 6 = 7 - x$$

$$\Rightarrow \qquad x = 7 - 6 = 1$$

3. $\dfrac{a+b-x}{c} + 1 + \dfrac{a+c-x}{b} + 1 + \dfrac{c+b-x}{a} + 1 + \dfrac{4x}{a+b+c} - 3 - 1 = 0$

$$\Rightarrow \quad \frac{a+b+c-x}{c} + \frac{a+b+c-x}{b} + \frac{a+b+c-x}{a}$$

$$+ \frac{4x-4a-4b-4c}{a+b+c} = 0$$

$$\Rightarrow (a+b+c-x)\left[\frac{1}{c} + \frac{1}{b} + \frac{1}{a} - \frac{4}{(a+b+c)}\right] = 0$$

$$\Rightarrow \quad a+b+c-x = 0$$

$$\Rightarrow \quad x = a+b+c$$

4. a,b,c are integers

$$a + b\sqrt{2} + c\sqrt{3} = 0$$

$$\Rightarrow \quad a + b\sqrt{2} = -c\sqrt{3}$$

Squaring both sides $2\sqrt{2}ab + a^2 + 2b^2 = 3c^2$

Comparing real parts and irrational parts

$a^2 + 2b^2 = 3c^2$, $ab = 0$

$$\Rightarrow \quad a = 0 \ or \ b = 0$$

case (i) If $a = 0$ then $3c^2 = 2b^2$

$$\Rightarrow \quad \sqrt{3}c = \pm\sqrt{2}b$$

$\sqrt{3}c$, $\sqrt{2}b$ are dissimilar ------, they will never be equal.

case (ii) If $b = 0$, then $3c^2 = a^2$

$$\Rightarrow \quad a = \pm\sqrt{3}c$$

If a, c are integers other than zeros one side becomes rational and other side becomes irrational.

$$\Rightarrow \quad a = 0, c = 0$$

$\therefore \ a = b = c = 0$.

5. $x^{\sqrt{x}} = x^{x/2}$

$\sqrt{x}$ is defined for $x \geq 0$

If $x = 0$, then $x^{\sqrt{x}} = 0^0$ is meaningless.

$\therefore \qquad\qquad x > 0$

If $x = 1$, then

L.H.S = $1^{\sqrt{1}} = 1$

R.H.S = $1^{1/2} = 1$

$\therefore \ x = 1$ is one root.

$$x^{\sqrt{x}} = x^{x/2}$$

$\Rightarrow \qquad \sqrt{x} = \dfrac{x}{2}$

$\Rightarrow \qquad 2\sqrt{x} = x$

$\Rightarrow \qquad x - 2\sqrt{x} = 0$

$\Rightarrow \qquad \sqrt{x}\left(\sqrt{x} - 2\right) = 0$

$\Rightarrow \qquad \sqrt{x} = 0 \qquad$ or $\qquad \sqrt{x} - 2 = 0$

$\Rightarrow \qquad x = 0 \qquad$ or $\qquad x = 4$

$\because \qquad x \neq 0$

$\therefore \qquad x = 4$ is the other root.

$\therefore \qquad x = \{1, 4\}$

6. From Pythagorean triplets, we have $3^2 + 4^2 = 5^2$

$\Rightarrow \qquad x = 2$ is a solution.

Let us find for other roots if at all possible.

$$3^x + 4^x = 5^x$$

$\Rightarrow \qquad \left(\dfrac{3}{5}\right)^x + \left(\dfrac{4}{5}\right)^x = 1$

If $x < 2$ then $\left(\dfrac{3}{5}\right)^x > \left(\dfrac{3}{5}\right)^2 \qquad\qquad \left(\because \ 0 < \dfrac{3}{5} < 1\right)$

Similarly, $\left(\dfrac{4}{5}\right)^x > \left(\dfrac{4}{5}\right)^2$

$\Rightarrow \qquad \left(\dfrac{3}{5}\right)^x + \left(\dfrac{4}{5}\right)^x > \left(\dfrac{3}{5}\right)^2 + \left(\dfrac{4}{5}\right)^2 = 1$

$\therefore \qquad \left(\dfrac{3}{5}\right)^x + \left(\dfrac{4}{5}\right)^x > 1$

$\therefore \qquad x < 2$ cannot be the solution.

If $x > 2$ then $\left(\dfrac{3}{5}\right)^x < \left(\dfrac{3}{5}\right)^2$

$\qquad\qquad \left(\dfrac{4}{5}\right)^x < \left(\dfrac{4}{5}\right)^2 \qquad\qquad \left(\because 0 < \dfrac{3}{5} < 1\right)$

$\Rightarrow \qquad \left(\dfrac{3}{5}\right)^x + \left(\dfrac{4}{5}\right)^x < \left(\dfrac{3}{5}\right)^2 + \left(\dfrac{4}{5}\right)^2 = 1$

$\therefore \qquad x > 2$ cannot be the solution.

$\therefore \qquad x = 2$ is the only solution.

7. Let $\qquad (x - ab) = c(a + b)$

$\qquad\qquad (x - ac) = b(a + c)$

$\qquad\qquad (x - bc) = a(b + c)$

$\Rightarrow \qquad x = ab + bc + ca$

8. $x = 1$ is satisfying both the given equations for all $a, b, c \in R$.

$\therefore \qquad x = 1$.

9. $f(x) = x^2 + 8x = x^2 + 2(x)(4) + 16 - 16$

$= (x + 4)^2 - 16 \geq -16$ for all $x \in R$.

$\therefore$ smallest value of $(x^2 + 8x) = -16$

10. $\sqrt{x+2\sqrt{x+\ldots\ldots\ldots+2\sqrt{x+2\sqrt{x+\ldots}}}} = x$

$\Rightarrow \quad \sqrt{x+2x} = x$

$\Rightarrow \quad \sqrt{3x} = x$

$\Rightarrow \quad 3x = x^2$

$\Rightarrow \quad x^2 - 3x = 0$

$\Rightarrow \quad x(x-3) = 0$

$\Rightarrow \quad x = 0 \quad \text{or} \quad x = 3$

11. $\dfrac{x+7+x+3}{(x+3)(x+7)} = \dfrac{x+9+x+1}{(x+1)(x+9)}$

$\Rightarrow (2x+10)(x+1)(x+9) = (2x+10)(x+3)(x+7)$

$\Rightarrow (2x+10)\left[(x+1)(x+9) - (x+3)(x+7)\right] = 0$

$\Rightarrow (2x+10)\left[x^2 +10x+9 - x^2 -10x -21\right] = 0$

$\Rightarrow (2x+10)(-12) = 0$

$\Rightarrow 2x+10 = 0$

$\Rightarrow x = -5$

12. By inspection, if $x = a+b$, then R.H.S. is not defined.

$\therefore \quad x = a+b$ is not a solution.

$\therefore \quad \dfrac{a}{x-a} + \dfrac{b}{x-b} = \dfrac{a}{(x-a-b)} + \dfrac{b}{(x-a-b)}$

$\dfrac{a}{x-a} - \dfrac{a}{(x-a-b)} = \dfrac{b}{(x-a-b)} - \dfrac{b}{(x-b)}$

$\Rightarrow a\left[\dfrac{x-a-b-x+a}{(x-a)(x-a-b)}\right] = b\left[\dfrac{x-b-x+a+b}{(x-a-b)(x-b)}\right]$

$$\Rightarrow -\frac{ab}{(x-a)(x-a-b)} = ab\left[\frac{1}{(x-a-b)(x-b)}\right]$$

$$\Rightarrow (x-a-b)(x-b) = -(x-a)(x-a-b)$$

$$\because x \neq a+b \Rightarrow -(x-a)(x-a-b) = (x-a-b)(x-b)$$

Dividing with $(x-a-b)$ throughout

$$\therefore \quad (x-b) = -(x-a)$$

$$\Rightarrow \quad 2x = a+b$$

$$\Rightarrow \quad x = \frac{a+b}{2}$$

13. $\dfrac{x-a}{b+c} - 1 + \dfrac{x-b}{c+a} - 1 + \dfrac{x-c}{a+b} - 1 = 0$

$$\Rightarrow \frac{x-a-b-c}{b+c} + \frac{x-a-b-c}{c+a} + \frac{x-a-b-c}{a+b} = 0$$

$$\Rightarrow (x-a-b-c)\left(\frac{1}{b+c} + \frac{1}{c+a} + \frac{1}{a+b}\right) = 0$$

$$\Rightarrow x-a-b-c = 0 \qquad \left[\text{if } \frac{1}{b+c} + \frac{1}{c+a} + \frac{1}{a+b} \neq 0\right]$$

$$\Rightarrow x = a+b+c$$

14. $\dfrac{x-a-b}{c} - 1 + \dfrac{x-b-c}{a} - 1 + \dfrac{x-c-a}{b} - 1 = 0$

$$\Rightarrow \frac{x-a-b-c}{c} + \frac{x-b-c-a}{a} + \frac{x-c-a-b}{b} = 0$$

$$\Rightarrow (x-a-b-c)\left(\frac{1}{c} + \frac{1}{a} + \frac{1}{b}\right) = 0$$

$$\Rightarrow (x-a-b-c)\left(\frac{ab+bc+ca}{abc}\right) = 0 \qquad \left(\begin{array}{l}\because ab+bc+ca \neq 0 \\ abc \neq 0\end{array}\right)$$

$\Rightarrow x - a - b - c = 0$

$\Rightarrow x = a + b + c$

15. $\sqrt{x-4} \geq 0 \Rightarrow x - 4 \geq 0 \Rightarrow x \geq 4$

$\therefore$ If $x < 4$ it will not satisfy the equation.

$\therefore \sqrt{3x+4} + \sqrt{x-4} = 2\sqrt{x}$

Squaring both sides

$3x + 4 + 2\sqrt{(3x+4)(x-4)} + x - 4 = 4x$

$\Rightarrow \quad 2\sqrt{(3x+4)(x-4)} = 0$

$\Rightarrow \quad (3x+4)(x-4) = 0$

$\Rightarrow \quad x = -\dfrac{4}{3} \quad or \quad x = 4$

$\because \quad x = -\dfrac{4}{3} < 4$ it is not a solution.

$\therefore \quad x = 4$ is the only solution.

16. $\dfrac{9-x-5}{x-4} + 3 = 0 \qquad \left[if \ x - 4 \neq 0 \right]$

$\Rightarrow \quad (9 - x - 5) + 3(x - 4) = 0$

$\Rightarrow \quad 4 - x + 3x - 12 = 0$

$\Rightarrow \quad 2x - 8 = 0$

$\Rightarrow \quad x = 4$ which is not possible.

$\therefore$ There is no solution.

17. Squaring both sides

$x + \sqrt{x+11} + x - \sqrt{x+11} + 2\sqrt{x^2 - (x+11)} = 16$

$\Rightarrow \quad x + \sqrt{x^2 - (x+11)} = 8$

$\Rightarrow \quad \sqrt{x^2 - (x+11)} = 8 - x$

Squaring both sides,

$$x^2 - (x+11) = 64 + x^2 - 16x$$

$$\Rightarrow \qquad 16x - x = 64 + 11$$

$$\Rightarrow \qquad 15x = 75$$

$$\Rightarrow \qquad x = 5$$

18. Squaring, $17 + x + (17 - x) - 2\sqrt{\left(17^2 - x^2\right)} = 4$

$$\Rightarrow \qquad 34 - 2\sqrt{17^2 - x^2} = 4$$

$$\Rightarrow \qquad 15 = \sqrt{17^2 - x^2}$$

Squaring,

$$\Rightarrow \qquad 225 = 289 - x^2$$

$$\Rightarrow \qquad x^2 = 289 - 225 = 64$$

$$\Rightarrow \qquad x = 8 \qquad \text{or} \qquad x = -8$$

Putting $x = -8$ in equation

$$\sqrt{17 - 8} - \sqrt{17 + 8} = 2$$

L.H.S $= \sqrt{9} - \sqrt{25} = 3 - 5 = -2$

R.H.S $= 2$ not satisfied.

$\therefore \quad x = 8$ is the solution.

19. $\sqrt{x - \sqrt{x - 2}} + \sqrt{x + \sqrt{x - 2}} = 2$

Squaring both sides

$$x - \sqrt{x - 2} + x + \sqrt{x - 2} + 2\sqrt{x^2 - (x - 2)} = 4$$

$$\Rightarrow \qquad 2x + 2\sqrt{x^2 - (x - 2)} = 4$$

$$\Rightarrow \qquad x + \sqrt{x^2 - (x - 2)} = 2$$

$$\Rightarrow \qquad \sqrt{x^2 - (x - 2)} = 2 - x$$

Squaring both sides,

$$x^2 - (x - 2) = 4 + x^2 - 4x$$

$\Rightarrow \quad 4x - x = 4 - 2$

$\Rightarrow \quad 3x = 2$

$\Rightarrow \quad x = \dfrac{2}{3}$

$\because \sqrt{x-2} \geq 0 \Rightarrow x - 2 \geq 0 \Rightarrow x \geq 2$

$\therefore x = \dfrac{2}{3}$ is not a solution.

$\therefore$ No solution.

20. $(x^2 - 4)\left(\sqrt{x+1}\right) = 0$

$\Rightarrow \quad (x+2)(x-2)\left(\sqrt{x+1}\right) = 0$

$\Rightarrow \quad x = -2 \ \ or \ \ x = 2 \ \ or \ \ \sqrt{x+1} = 0 \Rightarrow x + 1 = 0 \Rightarrow x = -1$

$\therefore \quad x = -1, -2, 2$

$\because \quad \sqrt{x+1} \geq 0 \Rightarrow x + 1 \geq 0 \Rightarrow x \geq -1$

$\therefore \quad x = -2$ is not a solution.

$\therefore \quad x = -1, 2$

21. $\sqrt[3]{1+\sqrt{x}} + \sqrt[3]{1-\sqrt{x}} = 2$

Cubing both sides,

$1 + \sqrt{x} + 1 - \sqrt{x} + 3\sqrt[3]{\left(1+\sqrt{x}\right)\left(1-\sqrt{x}\right)}\left(\sqrt[3]{1+\sqrt{x}} + \sqrt[3]{1-\sqrt{x}}\right) = 8$

$\left(\because \sqrt[3]{1+\sqrt{x}} + \sqrt[3]{1-\sqrt{x}} = 2\right)$

$\Rightarrow \quad 2 + 3\sqrt[3]{(1-x)}\,(2) = 8$

$\Rightarrow \quad 6\sqrt[3]{(1-x)} = 8 - 2 = 6$

$\Rightarrow \quad \sqrt[3]{(1-x)} = 1$

cubing both sides,

$\Rightarrow \qquad 1 - x = 1$

$\Rightarrow \qquad x = 0$ which satisfies the given equation.

22. $\sqrt[3]{x+1} + \sqrt[3]{x+2} + \sqrt[3]{x+3} = 0$

($\because a + b + c = 0$, *then* $a^3 + b^3 + c^3 = 3abc$)

$x + 3 + x + 1 + x + 2 = 3 \sqrt[3]{(x+1)(x+2)(x+3)}$

$3(x+2) = 3\sqrt[3]{(x+1)(x+2)(x+3)}$

$\Rightarrow \qquad x + 2 = \sqrt[3]{(x+1)(x+2)(x+3)}$

Cubing both sides

$\Rightarrow \qquad x^3 + 2^3 + 6x(x+2) = (x+1)(x+2)(x+3)$

$\Rightarrow \qquad x^3 + 6x^2 + 12x + 8 = (x^2 + 3x + 2)(x+3)$

$\Rightarrow \qquad x^3 + 6x^2 + 12x + 8 = x^3 + 6x^2 + 11x + 6$

$\Rightarrow \qquad x = 6 - 8 = -2$ which satisfies the given equation.

$\therefore \qquad x = -2 .$

23. $\sqrt[3]{x} + \dfrac{1}{\sqrt[3]{x}} = 3$

Cubing both sides,

$x + \dfrac{1}{x} + 3 \sqrt[3]{x} \cdot \dfrac{1}{\sqrt[3]{x}} \left(\sqrt[3]{x} + \dfrac{1}{\sqrt[3]{x}} \right) = 27 \qquad\qquad \left(\because \sqrt[3]{x} + \dfrac{1}{\sqrt[3]{x}} = 3 \right)$

$\Rightarrow \qquad x + \dfrac{1}{x} + 3(3) = 27$

$\Rightarrow \qquad x + \dfrac{1}{x} = 18$

Cubing both sides

$\left(x + \dfrac{1}{x} \right)^3 = 18^3$

$$x^3 + \frac{1}{x^3} + 3\left(x + \frac{1}{x}\right) = 18^3 \qquad \left(\because \ x + \frac{1}{x} = 18\right)$$

$$\Rightarrow \qquad x^3 + \frac{1}{x^3} = 18^3 - 3 \times 18 \qquad \left(\because \ x + \frac{1}{x} = 18\right)$$

$$\Rightarrow \qquad x^3 + \frac{1}{x^3} = 18(18^2 - 3) = 18\,(324 - 3)$$

$$= 18 \times 321 = 5778$$

$$\therefore \qquad x^3 + \frac{1}{x^3} = 5778$$

24. $$\frac{x}{a+b} + \frac{a+b}{a+b} = \frac{x}{a-b} + \frac{a-b}{a+b}$$

$$\Rightarrow \qquad \frac{x}{a+b} - \frac{x}{(a-b)} = \frac{a-b}{a+b} - \frac{a+b}{a+b}$$

$$\Rightarrow \qquad \frac{ax - bx - ax - bx}{a^2 - b^2} = \frac{a - b - a - b}{(a+b)}$$

$$\Rightarrow \qquad \frac{-2bx}{(a+b)(a-b)} = \frac{-2b}{(a+b)}$$

$$\Rightarrow \qquad \frac{x}{a-b} = 1$$

$$\Rightarrow \qquad x = a - b$$

25. $$\frac{m(x+a)}{x+b} - m + \frac{n(x+b)}{x+a} - n = 0$$

$$\Rightarrow \qquad \frac{mx + ma - mx - mb}{x+b} + \frac{nx + nb - nx - na}{x+a} = 0$$

$$\Rightarrow \qquad \frac{m(a-b)}{x+b} + \frac{n(b-a)}{x+a} = 0$$

$$\Rightarrow \qquad (a-b)\left(\frac{m}{x+b} - \frac{n}{x+a}\right) = 0$$

35

$$\Rightarrow \quad \frac{m}{x+b} = \frac{n}{x+a} \qquad (\because \ a - b \neq 0)$$

$$\Rightarrow \quad mx + ma = nx + nb \Rightarrow (m-n)x = nb - ma$$

$$\Rightarrow \quad x = \frac{nb - ma}{m - n}$$

26. $$\frac{\left(\sqrt{x+3} - \sqrt{x+2}\right)^2}{(x+3)-(x+2)} = \frac{4}{5} \qquad \text{(Rationalise)}$$

$$\Rightarrow \quad \left(\sqrt{x+3} - \sqrt{x+2}\right)^2 = \frac{4}{5}$$

$$\Rightarrow \quad x + 3 + x + 2 - 2\sqrt{(x+3)(x+2)} = \frac{4}{5}$$

$$\Rightarrow \quad 2x + 5 - 2\sqrt{(x+3)(x+2)} = \frac{4}{5}$$

$$\Rightarrow \quad 10x + 25 - 10\sqrt{(x+3)(x+2)} = 4$$

$$\Rightarrow \quad (10x + 21) = 10\sqrt{(x+3)(x+2)}$$

Squaring both sides

$$100x^2 + 441 + 420x = 100(x^2 + 5x + 6) = 100x^2 + 500x + 600$$

$$\Rightarrow \quad 80x = -159$$

$$\Rightarrow \quad x = -\frac{159}{80}$$

EXERCISE - 2 [Time: 3hrs]

1. Solve

$$(x+a)^2 + (x+b)^2 + (x+c)^2 = (x-3a)^2 + (x-3b)^2 + (x-3c)^2$$

2. Solve $\dfrac{2x}{1 + \dfrac{1}{1 + \dfrac{x}{1-x}}} = 1$

3. Solve $\dfrac{x - a^3}{b^2 - bc + c^2} + \dfrac{x - b^3}{c^2 - ca + a^2} + \dfrac{x - c^3}{a^2 - ab + b^2} = 2(a+b+c)$

4. If $a \neq b$, then solve $\dfrac{x+a}{x+b} = \left(\dfrac{2x+a+c}{2x+b+c}\right)^2$

5. Solve: $(x-10)^3 + (x-30)^3 = 2(x-20)^3$

6. Solve: $\dfrac{4x^2+7}{2x-1} + \dfrac{6x^2-8x+11}{3x-1} = \dfrac{4x^2+3x+6}{x+1}$

7. Solve: $\dfrac{1}{1-\sqrt{1-x}} - \dfrac{1}{1+\sqrt{1-x}} = \dfrac{\sqrt{2}}{x}$

8. Solve: $\dfrac{ax+b}{a+bx} = \dfrac{a^2x^2-b^2}{b^2x^2-a^2}$ if $a \neq b$

9. Solve: $\sqrt[3]{a-x} + \sqrt[3]{b-x} = \sqrt[3]{a+b-2x}$ by inspection.

EXERCISE - 2
Answers

1. $\dfrac{a^2+b^2+c^2}{a+b+c}$

2. $\dfrac{2}{3}$

3. $a^3+b^3+c^3$

4. $\dfrac{ab-c^2}{2c-a-b}$

5. 20

6. $\dfrac{2}{5}$

7. $\dfrac{1}{2}$

8. $-\dfrac{b}{a}, -1$

9. $a, b, a+\dfrac{b}{2}$

BLANK PAGE LEFT INTENTIONALLY

3. QUADRATIC EQUATION

1. Quadratic Equation

The standard form of a quadratic equation is:

$$ax^2 + bx + c = 0$$

where a, b, c are real numbers and $a \neq 0$

2. Roots of a quadratic equation

roots of a quadratic equation

$$ax^2 + bx + c = 0 \qquad (a \neq 0,\ a, b, c \in R)$$

are given by

$$\alpha = \frac{-b + \sqrt{b^2 - 4ac}}{2a} \ ; \ \beta = \frac{-b - \sqrt{b^2 - 4ac}}{2a}$$

* sum of the roots $= \alpha + \beta = -\dfrac{b}{a}$

* Product of roots $= \alpha\beta = \dfrac{c}{a}$

* factorised form of $ax^2 + bx + c = a(x - \alpha)(x - \beta)$

* If S be the sum and P be the product of roots, then quadratic equation is:

$$x^2 - sx + p = 0$$

3. Nature of roots of a quadratic equation

Nature of roots of a quadratic equation

$$ax^2 + bx + c = 0$$

means whether the roots are real or complex.

By analysing the expression

$$D = b^2 - 4ac$$

(D called as discriminant), one can get an idea about the nature of the roots as follows:

(i) (a) If $D < 0$ $\qquad (b^2 - 4ac < 0)$

then the roots of the quadratic equation are non-real or complex roots

(b) If $D = 0$ $\qquad (b^2 - 4ac = 0)$

then the roots are real and equal.

Equal roots $\alpha = \beta = -\dfrac{b}{2a}$

(c) If $D > 0$ $(b^2 - 4ac > 0)$

then the roots are real and unequal.

(ii)   If D i.e., $(b^2 - 4ac)$ is a perfect square and a, b and c are rational, then the roots are rational.

(iii) If D i.e., $(b^2 - 4ac)$ is not a perfect square and a, b and c are rational, then the roots are of the form

$m + \sqrt{n}$ and $m - \sqrt{n}$.

(iv)   If $D < 0$ i.e., $(b^2 - 4ac < 0)$, and the coefficients a, b and c are real then the roots are complex conjugate of each other i.e., the roots are of the form

$p + iq$ and $p - iq$ ($p, q \in R$ and $i = \sqrt{-1}$).

(v)   If $a = 1$, $b, c \in I$ and the roots are rational numbers, then the roots must be integer.

(vi)   If a quadratic equation in x has more than two roots, then it is an identity in x (i.e. true for all real values of x) and $a = b = c = 0$.

4.   Condition for common root(s)

consider two quadratic equations:

$a_1 x^2 + b_1 x + c_1 = 0$ and $a_2 x^2 + b_2 x + c_2 = 0$

(a) For two common roots:

In such a case, two equations should be identical. For that, the ratio of coefficients of x^2, x and x^0 must be same,

i.e., $\dfrac{a_1}{a_2} = \dfrac{b_1}{b_2} = \dfrac{c_1}{c_2}$

(b) For one common root:

Let α be the common root of two equations. So α should satisfy the two equations.

$\Rightarrow$ $a_1 \alpha^2 + b_1 \alpha + c_1 = 0$ and $a_2 \alpha^2 + b_2 \alpha + c_2 = 0$

Solving the two equations by using cross multiplication method

$$\Rightarrow \qquad \frac{\alpha^2}{b_1 c_2 - b_2 c_1} = \frac{-\alpha}{a_1 c_2 - a_2 c_1} = \frac{1}{a_1 b_2 - a_2 b_1}$$

$$\Rightarrow \qquad \alpha = \frac{a_2 c_1 - a_1 c_2}{a_1 b_2 - a_2 b_1}$$

$$\alpha^2 = \frac{b_1 c_2 - b_2 c_1}{a_1 b_2 - a_2 b_1}$$

$\Rightarrow \qquad (b_1 c_2 - b_2 c_1)(a_1 b_2 - a_2 b_1) = (a_2 c_1 - a_1 c_2)^2$. This is the condition for one root of two quadratic equations to be common.

Note: To find the common root between the two equations, make the coefficient of α^2 common and then subtract the two equations.

5. Some more result on roots of quadratic equation.

(i) Both roots of $f(x) = 0$ are negative, if sum of the roots < 0, product of the roots > 0 and $D \geq 0$.

i.e., $\qquad -\dfrac{b}{a} < 0, \ \dfrac{c}{a} > 0, \ b^2 - 4ac \geq 0$

(ii) Both roots of $f(x) = 0$ are positive, if sum of the roots > 0, product of the roots > 0 and $D \geq 0$

i.e., $\qquad -\dfrac{b}{a} > 0, \dfrac{c}{a} > 0, \ b^2 - 4ac \geq 0$

(iii) Roots of $f(x) = 0$ are opposite in sign, if product of the roots < 0,

i.e., $\qquad \dfrac{c}{a} < 0$.

6. Symmetric third-degree equations.

An algebraic third degree equation is said to be symmetric if it has the form

$$ax^3 + bx^2 + bx + a = 0 \qquad\qquad (a \neq 0)$$

The equivalent equation to the above equation is

$$(x+1)\left[ax^2 + (b-a)x + a \right] = 0 \qquad\qquad (a \neq 0)$$

Now we can solve

$\because \qquad x+1=0 \quad \Rightarrow \quad x=-1$

or $\qquad ax^2 + (b-a)x + a = 0$

This equation is a quadratic.

7. Symmetric fourth-degree equations.

    An algebraic fourth-degree equation is said to be symmetric if it has the form

    $$ax^4 + bx^3 + cx^2 + bx + a = 0 \qquad (a \neq 0)$$

    The equivalent equation to the above equation is

    $$\left(x^2 + \frac{b}{2a}x + 1\right)^2 - \frac{b^2 - 4a(c-2a)}{4a^2}x^2 = 0 \qquad (a \neq 0) \qquad \ldots (i)$$

    Let $\qquad M = b^2 - 4a(c-2a)$

    Case (i) $M < 0$, there is no real solution to the given equation.

    Case (ii) $M = 0$, then $\left(x^2 + \frac{b}{2a}x + 1\right)^2 = 0$

    $\Rightarrow \qquad \left(x^2 + \frac{b}{2a}x + 1\right) = 0$

    The roots of the fourth-degree symmetric equation coincides with the set of roots of the quadratic equation

    $$x^2 + \frac{b}{2a}x + 1 = 0 \qquad (a \neq 0)$$

    Case (iii): $M > 0$, then

    The equivalent equation (i) can be written as

    $$\left[x^2 + \frac{b+\sqrt{b^2-4a(c-2a)}}{2a}x + 1\right]\left[x^2 + \frac{b-\sqrt{b^2-4a(c-2a)}}{2a} = 0\right] \, (a \neq 0)$$

    which are two quadratic equations.

EXERCISE - 1

1. If α and β are the roots of equation $ax^2 + bx + c = 0$, find the value

of the following expressions:

(i) $\alpha^2 + \beta^2$ (ii) $\alpha^3 + \beta^3$ (iii) $\alpha^4 + \beta^4$

(iv) $(\alpha - \beta)^2$ (v) $\alpha^4 - \beta^4$

2. If α and β are the roots of equation $ax^2 + bx + c = 0$, form an equation whose roots are:

(i) $\alpha + \dfrac{1}{\beta}, \alpha + \dfrac{1}{\beta}, \beta + \dfrac{1}{\alpha}$ (ii) $\dfrac{1}{\alpha + \beta}, \dfrac{1}{\alpha} + \dfrac{1}{\beta}$

3. Form an equation whose roots are squares of the sum and the difference of the roots of the equation.

$2x^2 + 2(m+n)x + m^2 + n^2 = 0$

4. Comment upon the nature of the following equation:

(i) $x^2 + (a+b)x - c^2 = 0$

(ii) $(a+b+c)x^2 - 2(a+b)x + (a+b-c) = 0$

(iii) $(b-c)x^2 + (c-a)x + (a-b) = 0$

5. What can you say about the roots of the following equations?

(i) $x^2 + 2(3a+5)x + 2(9a^2 + 25) = 0$

(ii) $(x-a)(x-b) + (x-b)(x-c) + (x-c)(x-a) = 0$

6. Find the values of k, so that the equations $2x^2 + kx - 5 = 0$ and $x^2 - 3x - 4 = 0$ many have one root in common.

7. If $ax^2 + bx + c = 0$ and $bx^2 + cx + a = 0$ have a root in common, find the relation between a, b and c.

8. If the equations $x^2 - ax + b = 0$ and $x^2 - cx + d = 0$ have one root in common and second equation has equal roots, prove that $ac = 2(b+d)$.

9. If α, β are the roots of $x^2 + px + q = 0$ and γ, δ are the roots of $x^2 + rx + s = 0$, evaluate the value of $(\alpha - \gamma)(\alpha - \delta)(\beta - \gamma)(\beta - \delta)$ in terms of p, q, r, s. Hence deduce the condition that the equations have a common root.

10. If the ratio of roots of the equation $x^2 + px + q = 0$ be equal to the

ratio of roots of the equation $x^2 + bx + c = 0$, then prove that $p^2 c = b^2 q$.

11. Find the roots of the equation

$2\log_x a + \log_{ax} a + 3\log_{a^2 x} a = 0$ if $a > 0$, $a \neq 1$.

12. Find the condition for the equation $\dfrac{1}{x} + \dfrac{1}{x+b} = \dfrac{1}{m} + \dfrac{1}{m+b}$ has real roots that are equal in magnitude but opposite in sign.

13. For what value of a does the equation $\log (x^2 + 2ax) = \log(8x - 6a - 3)$ have only one solution?

14. Find the real roots of the equation

$$\sqrt{x + 3 - 4\sqrt{x-1}} + \sqrt{x + 8 - 6\sqrt{x-1}} = 1$$

15. Solve the equation: $\sqrt{x-2} + \sqrt{4-x} = \sqrt{6-x}$

16. Solve the equation

$$\log_{x^2 + 6x + 8} \log_{2x^2 + 2x + 3}(x^2 - 2x) = 0$$

17. Solve the following equation for x :

$$\log_{2x+3}(6x^2 + 23x + 21) + \log_{3x+7}(4x^2 + 12x + 9) = 4$$

18. Solve $\dfrac{x-a}{x-b} + \dfrac{x-b}{x-a} = \dfrac{a}{b} + \dfrac{b}{a}$

19. Find the roots of the equation $x^3 + 4x^2 + 4x + 1 = 0$

20. Solve the equation $x^4 + x^3 - x^2 + x + 1 = 0$

21. Solve the equation

$$x^4 - x^2 - 6 = 0$$

22. Solve: $x^4 - 10x^3 + 26x^2 - 10x + 1 = 0$

23. Solve the equation $4x - 3.2^x + 2 = 0$

24. Solve the equation:

$3^{\log_3 (x^2 - 4x + 3)} = x - 3$

25. Solve the equation

$$\frac{(a-x)\sqrt{a-x}-(b-x)\sqrt{x-b}}{\sqrt{a-x}+\sqrt{x-b}}=a-b$$

26. Solve the equation

$$\sqrt{4a+b-5x}+\sqrt{4b+a-5x}-3\sqrt{a+b-2x}=0$$

27. Prove that the roots of the equation

$$(x-a)(x-c)+\lambda(x-b)(x-d)=0$$

are real for any λ if $a<b<c<d$.

28. Show that the roots of the equation

$$(x-a)(x-b)+(x-a)(x-c)+(x-b)(x-c)=0$$ are always real.

29. Prove that at least one of the equations

$$x^2+px+q=0$$

$$x^2+p_1x+q_1=0$$

has real roots if $p_1p=2(q_1+q)$

30. Prove that at least one of the roots of equation

$$a(x-b)(x-c)+b(x-a)(x-c)+c(x-a)(x-b)=0$$ are always real.

31. Find the values of p and q for which the roots of the equation

$$x^2+px+q=0$$ are equal to p and q.

32. Solve $\left(x+\frac{1}{x}\right)^2-\frac{3}{2}\left(x-\frac{1}{x}\right)=4$, when $x\neq 0$

33. Solve $x^4-2x^3-x^2-2x+1=0$

34. Solve $(x-1)(x-3)(x-5)(x-7)=9$

35. Find 'k' if one root of $kx^2-14x+8=0$ may be six times other.

36. If α,β are the roots of the equation $x^2+x-1=0$ then equation whose roots are $\alpha+2,\beta+2$.

37. If $2+\sqrt{3}$ is one root of $x^2+px+q=0$ then find $'p','q'$

38. Solve $\sqrt{2x-6}+\sqrt{x+4}=5$

39. If the equations $(a^2-4a+3)x^2+(a-1)x+a^2-1=0$ has infinite roots

then find $'a'$.

40.   Equations $px^2 + qx + r = 0$ and $qx^2 - 2\sqrt{pr}x + q = 0$ have real roots then show that $'p','q','r'$ are in G.P.

41.   If $'a'$ and $'b'$ are the roots of $11x^2 - 4x - 2 = 0$ then compute the product $(1 + a + a^2 + ...\infty)(1 + b + b^2 + ... \infty)$

42.   If the quadratic $ax^2 + bx + c = 0(a \neq 0) a, b, c$ are integers, has natural numbers as it roots, then S.T. a divides $'b'$ & $'c'$.

       (a) ac can be expressed as the sum of two squares of natural numbers

       (b) $'a'$ divides $'b'$ & $'c'$

       (c) $'b'$ divides $'c'$ & $'a'$

       (d) $'c'$ divides $'a'$ & $'b'$

       (e) None of these

43.   The value of $'a' \in R$ for which the equation

       $(1 + a^2)x^2 + 2(x - a)(1 + ax) + 1 = 0$ has no real roots.

44.   Let α, β are the roots of the quadratic equation $x^2 + ax + b = 0$ and γ, δ be the roots of the equation $x^2 - ax + b - 2 = 0$.

       Given that $1/\alpha + 1/\beta + 1/\gamma + 1/\delta = 5/12$ and $\alpha\beta\gamma\delta = 24$. Find the value of the coefficient $'a'$.

45.   Solve $3x^2 - 4\sqrt{3x^2 - 4x + 1} = 4x - 4$

Exercise - 1
Solution

1.   (i) $\alpha + \beta = -\dfrac{b}{a}$ and $\alpha\beta = \dfrac{c}{a}$

$$\alpha^2 + \beta^2 = (\alpha + \beta)^2 - 2\alpha\beta = \left(\dfrac{-b}{a}\right)^2 - \dfrac{2c}{a}$$

$$= \dfrac{b^2 - 2ac}{a^2}$$

(ii) $\alpha^3 + \beta^3 = (\alpha+\beta)^3 - 3\alpha\beta(\alpha+\beta)$

$$= \left(-\frac{b}{a}\right)^3 - 3\left(\frac{c}{a}\right)\left(-\frac{b}{a}\right) = \frac{-b^3 + 3abc}{a^3}$$

(iii) $\alpha^4 + \beta^4 = (\alpha^2+\beta^2) - 2\alpha^2\beta^2$

$$= \left(\frac{b^2-2ac}{a^2}\right)^2 - 2\left(\frac{c}{a}\right)^2 = \frac{(b^2-2ac)^2 - 2c^2a^2}{a^4}$$

(iv) $(\alpha-\beta)^2 = (\alpha+\beta)^2 - 4\alpha\beta = \frac{b^2}{a^2} - \frac{4c}{a} = \frac{b^2-4ac}{a^2}$

(v) $\alpha^4 - \beta^4 = (\alpha^2+\beta^2)(\alpha+\beta)(\alpha-\beta)$

$$= \left(\frac{b^2-2ac}{a^2}\right)\left(-\frac{b}{a}\right)\left(\pm\sqrt{\frac{b^2-4ac}{a^2}}\right)$$

$$= \pm\frac{b}{a^4}(b^2-2ac)\sqrt{b^2-4ac}$$

2.　(i) $\text{Sum(s)} = \left(\alpha+\frac{1}{\beta}\right) + \left(\beta+\frac{1}{\alpha}\right)$

$$= (\alpha+\beta) + \frac{\alpha+\beta}{\alpha\beta} = \frac{-b(a+c)}{ac}$$

$\text{Product (p)} = \left(\alpha+\frac{1}{\beta}\right)\left(\beta+\frac{1}{\alpha}\right) = \alpha\beta + \frac{1}{\alpha\beta} + 2$

$$= \frac{(c+a)^2}{ca}$$

The equation is $= x^2 - sx + p = 0 \dfrac{x-\mu}{\sigma}$

$$\Rightarrow x^2 - \left(\frac{-b(a+c)}{ac}\right)x + \frac{(c+a)^2}{ac} = 0$$

$$\Rightarrow ac\,x^2 + b(c+a)x + (c+a)^2 = 0$$

47

(ii) Sum(s) $= \left(\dfrac{1}{\alpha+\beta}\right) + \left(\dfrac{1}{\alpha}+\dfrac{1}{\beta}\right) = \left(\dfrac{1}{\alpha+\beta}\right) + \dfrac{(\alpha+\beta)}{\alpha\beta}$

$$= -\dfrac{(ac+b)^2}{bc}$$

Product (p) $= \left(\dfrac{1}{\alpha+\beta}\right)\left(\dfrac{1}{\alpha}+\dfrac{1}{\beta}\right) = \left(\dfrac{1}{\alpha\beta}\right) = \dfrac{a}{c}$

The equation is: $x^2 - sx + p = 0$

$\Rightarrow \qquad x^2 - \left(-\dfrac{(ac+b^2)}{bc}\right)x + \dfrac{a}{c} = 0$

$\Rightarrow \qquad bcx^2 + (ac+b^2)x + ab = 0$ is the required equation.

3.   Let α, β are the roots of given equation.

$\Rightarrow \qquad \alpha+\beta = -(m+n)$ and $\alpha\beta = \dfrac{(m^2+n^2)}{2}$

We have to get the equation whose roots are $(\alpha+\beta)^2$ and $(\alpha-\beta)^2$

Sum (s) $= (\alpha+\beta)^2 + (\alpha-\beta)^2$

$$= 2(\alpha^2+\beta^2) = 2\left[(\alpha+\beta)^2 - 2\alpha\beta\right] = 4mn$$

product (p) $= (\alpha+\beta)^2 . (\alpha-\beta)^2$

$$= (\alpha+\beta)^2 . \left[(\alpha+\beta)^2 - 4\alpha\beta\right]$$

$p = (m+n)^2\left[(m+n)^2 - 2(m^2+n^2)\right] = -(m^2-n^2)^2$

The equation is: $x^2 - Sx + p = 0$

$\therefore$ The required equation is

$$x^2 - 4mnx - (m^2 - n^2)^2 = 0$$

4.   (i) Discriminant (D)

$D = (a+b)^2 - 4(1)(-c^2) = (a+b)^2 + 4c^2$

$\Rightarrow \qquad D \geq 0$

$\therefore$ The roots are real.

(ii) $D = 4(a+b)^2 - 4(a+b+c)(a+b-c)$

$\quad = 4\left[(a+b)^2 - \left\{(a+b)^2 - c^2\right\}\right]$

$\quad = 4\left\{(a+b)^2 - (a+b)^2 + c^2\right\} = 4c^2 = (2c)^2$

$\Rightarrow \quad D \geq 0$ and also a perfect square hence the roots are rational.

(iii) $D = (c-a)^2 - 4(b-c)(a-b)$

$\quad\quad\quad = c^2 + a^2 + (2b)^2 - 4ab - 4bc + 2ac$

$\quad\quad\quad = c^2 + a^2 + (2b)^2 - 4ab - 4bc + 2ac$

$\quad\quad\quad = (c+a-2b)^2$

$\Rightarrow \quad D \geq 0$ and also a perfect square.

$\therefore$ The roots are rational.

5. (i) $D = 4(3a+5)^2 - 8(9a^2 + 25) = -4(3a-5)^2$

$\Rightarrow D \leq 0$, so the roots are non real if $a \neq 5/3$ and real and equal if

$a = \dfrac{5}{3}$.

(ii) Simplifying the given equation :

$3x^2 - 2(a+b+c)x + (ab+bc+ca) = 0$

$D = 4(a+b+c)^2 - 12(ab+bc+ca)$

$\quad = 4(a^2+b^2+c^2 - ab - bc - ca)$

$\quad = 2\left[(a-b)^2 + (b-c)^2 + (c-a)^2\right] \quad \begin{bmatrix} \because (a^2+b^2+c^2 - ab - bc - ca) \\ = \dfrac{1}{2}\left[(a-b)^2 + (b-c)^2 + (c-a)^2\right] \end{bmatrix}$

$\Rightarrow \quad\quad D \geq 0$, so the roots are real.

Note: If $D = 0$, then $\left[(a-b)^2 + (b-c)^2 + (c-a)^2\right] = 0$

$\Rightarrow \quad\quad a = b = c$

$\Rightarrow \quad\quad$ If $a = b = c$, then the roots are equal.

6. Let α be the common root of two equations.

49

$$\therefore \qquad 2\alpha^2 + k\alpha - 5 = 0$$

$$\alpha^2 - 3\alpha - 4 = 0$$

Solving the two equations

$$\frac{\alpha^2}{-4k-15} = \frac{-\alpha}{-8+5} = \frac{1}{-6-k}$$

$$\Rightarrow \qquad (-3)^2 = (4k+15)(6+k)$$

$$\Rightarrow \qquad 4k^2 + 39k + 81 = 0$$

$$\Rightarrow \qquad k = -3 \quad or \quad k = -\frac{27}{4}$$

7.   Using the condition for common root, we have

$$(a^2 - bc)^2 = (ba - c^2)(ac - b^2)$$

$$\Rightarrow \qquad a^4 + b^2c^2 - 2a^2bc = a^2bc - b^3a - ac^3 + b^2c^2$$

$$\Rightarrow \qquad a(a^3 + b^3 + c^3 - 3abc) = 0$$

$$\Rightarrow \qquad a = 0 \quad or \quad a^3 + b^3 + c^3 - 3abc = 0$$

This is the relation between a, b and c. From second relation, we also have the relation $a + b + c = 0$

8.   The equation $x^2 - cx + d = 0$ has equal root.

$$\therefore \qquad D = 0 \Rightarrow D = c^2 - 4d = 0 \qquad\qquad\qquad \text{... (i)}$$

and the equal roots are

$$x = \frac{-(c-) \pm \sqrt{c^2 - 4d}}{2(1)} = \frac{c \pm 0}{2} = \frac{c}{2}$$

$$\Rightarrow \qquad x = \frac{c}{2} \text{ is the equal root of this equation.}$$

$$\therefore \quad \text{This is the common root of the both the equations.}$$

$$\therefore \quad x = \frac{c}{2} \text{ will satisfy the first equation.}$$

$$\Rightarrow \qquad \frac{c^2}{4} - a\left(\frac{c}{2}\right) + b = 0$$

$$\Rightarrow \quad c^2 + 4b = 2ac$$

$$\Rightarrow \quad 4d + 4b = 2ac \qquad \left(\because c^2 = 4d \ \ from\,(i)\right)$$

$$\Rightarrow \quad 2(d + b) = ac$$

$$\therefore \quad ac = 2(b + d)$$

9.   Roots of $x^2 + px + q = 0$ are α, β

     Roots of $x^2 + rx + s = 0$ are γ, δ

$$\therefore \quad \alpha + \beta = -p \qquad and \qquad \alpha\beta = q$$

$$\gamma + \delta = -r \qquad and \qquad \gamma\delta = s$$

$$\therefore \quad (\alpha - \gamma)(\alpha - \delta)(\beta - \gamma)(\beta - \delta)$$

$$= \left[\alpha^2 - (\gamma + \delta)\alpha + \gamma\delta\right]\left[\beta^2 - (\gamma + \delta)\beta + \gamma\delta\right]$$

$$= (\alpha^2 + r\alpha + s)\,(\beta^2 + r\beta + s)$$

$$(\because \alpha^2 + p\alpha + q = 0 \ \ and \ \ \beta^2 + p\beta + q = 0)$$

$$= (-p\alpha - q + r\alpha + s)(-p\beta - q + r\beta + s)$$

$$= \left[(r - p)\alpha + (s - q)\right]\left[(r - p)\beta + (s - q)\right]$$

$$= (r - p)^2 \alpha\beta + (s - v)^2 + (s - v)(r - p)(\alpha + \beta)$$

$$= (r - p)^2 q + (s - q)^2 + (s - q)(r - p)(-p)$$

$$= (q - p)[rq - pq - ps + pq] + (s - v)^2$$

$$= (r - p)(rq - ps) + (s - v)^2$$

If the equation have a common root then either

$$\alpha = r \qquad or \qquad \alpha = s \qquad or \qquad \beta = r \qquad or \qquad p = s$$

i.e. $(\alpha - r)(\alpha - s)(p - r)(\beta - s) = 0$

$$\Rightarrow \quad (s - q)^2 + (r - p)(rq - ps) = 0$$

$$\Rightarrow \quad (s - q)^2 = (r - p)(ps - qr)$$

10. $\dfrac{\alpha}{\beta} = \dfrac{\gamma}{\delta} \quad \Rightarrow \quad \dfrac{(\alpha + \beta)^2}{(\alpha - \beta)^2} = \dfrac{(\gamma + \delta)^2}{(\gamma - \delta)^2}$

$$\Rightarrow \quad \frac{(\alpha+\beta)^2}{(\alpha+\beta)^2-(\alpha-\beta)^2}=\frac{(\gamma+\delta)^2}{(\gamma+\delta)^2-(\gamma-\delta)^2}$$

$$\Rightarrow \quad \frac{(\alpha+\beta)^2}{4\alpha\beta}=\frac{(\gamma+\delta)^2}{4\gamma\delta}$$

$$\Rightarrow \quad \frac{p^2}{4q}=\frac{b^2}{4c}$$

$$\Rightarrow \quad p^2 c = b^2 q$$

11. The given equation can be written as

$$2\frac{\log a}{\log x}+\frac{\log a}{\log ax}+\frac{3\log a}{\log a^2 x}=0 \qquad (\because a>0 \; and \; a\neq 1, \; \log a \neq 0)$$

$$\Rightarrow \quad \frac{2}{y}+\frac{1}{b+y}+\frac{3}{2b+y}=0 \qquad (\text{where } b=\log a \; and \; y=\log x)$$

$$\Rightarrow \quad 2(b+y)(2b+y)+y(2b+y)+3y(b+y)=0$$

$$\Rightarrow \quad 4b^2+11by+6y^2=0$$

which is quadratic in y.

$$\therefore \quad y=\frac{-11b\pm\sqrt{121b^2-96b^2}}{12}$$

$$\Rightarrow \quad y=\frac{-4b}{3}, \; -\frac{b}{2} \qquad (\because y=\log x \; and \; b=\log a)$$

$$\Rightarrow \quad \log x=-\frac{4}{3}\log a \quad or \qquad \log x=-\frac{\log a}{2}$$

$$\Rightarrow \quad x=a^{-\frac{4}{3}} \; or \qquad x=a^{-\frac{1}{2}}$$

12. From the given equation $x=m$ is a root.

$\therefore$ The other root must be –m

$$\frac{1}{-m}+\frac{1}{-m+b}=\frac{1}{m}+\frac{1}{m+b}$$

$$\Rightarrow \quad \frac{1}{b-m}-\frac{1}{b+m}=\frac{2}{m}$$

$$\Rightarrow \qquad \frac{b+m-b+m}{b^2-m^2} = \frac{2}{m}$$

$$\Rightarrow \qquad 2m^2 = 2b^2 - 2m^2$$

$$\Rightarrow \qquad 2m^2 = b^2$$

13. $\log(x^2 + 2ax) = \log(8x - 6a - 3)$

$$\Rightarrow \qquad x^2 + 2ax = 8x - 6a - 3$$

$$\Rightarrow \qquad x^2 + (2a-8)x + 3(2a+1) = 0$$

$$D = (2a-8)^2 - 4 \times 3(2a+1)$$

For one solution to exist $D = 0$

$$(a-4)^2 - 3(2a+1) = 0$$

$$\Rightarrow \qquad a^2 - 14a + 13 = 0$$

$$\Rightarrow \qquad (a-1)(a-13) = 0$$

$$\Rightarrow \qquad a = 1, 13$$

14. Let $x - 1 = t^2$

$$\sqrt{x+3-4\sqrt{x-1}} + \sqrt{x+8-6\sqrt{x-1}} = 1$$

$$\Rightarrow \qquad \sqrt{t^2+4-4t} + \sqrt{t^2+9-6t} = 1$$

$$\Rightarrow \qquad \sqrt{(t-2)^2} + \sqrt{(t-3)^2} = 1$$

$$\Rightarrow \qquad |t-2| + |t-3| = 1$$

Case (i) $\qquad t \le 2$

$$2 - t + 3 - t = 1$$

$$\Rightarrow \qquad 5 - 2t = 1$$

$$\Rightarrow \qquad t = 2$$

$$\Rightarrow \qquad x - 1 = 4$$

$$\Rightarrow \qquad x = 5 \qquad\qquad \dots \text{(i)}$$

Case (ii) $\qquad 2 < t < 3$

$$t - 2 + 3 - t = 1$$

$1 = 1 \Rightarrow \quad true \; \forall \; t \in (2,3)$

$\Rightarrow \qquad 4 < t^2 < 9$

$\Rightarrow \qquad 5 < x < 10 \qquad \qquad \qquad$ ... (ii)

Case (iii) $\qquad \quad t \geq 3$

$\qquad \qquad t - 2 + t - 3 = 1$

$\Rightarrow \qquad 2t = 6 \qquad \qquad \Rightarrow \qquad t = 3$

$\Rightarrow \qquad t^2 = 9$

$\Rightarrow \qquad x - 1 = 9 \quad \Rightarrow \quad x = 10 \qquad \qquad$ ... (iii)

Combining (i), (ii) and (iii); $x \in [5, 10]$

15. $\quad \sqrt{x-2} + \sqrt{4-x} = \sqrt{6-x} \qquad \qquad$ ... (i)

On squaring both sides

$\Rightarrow \qquad (x-2) + (4-x) + 2\sqrt{(x-2)(4-x)} = 6-x$

$\Rightarrow \qquad 2 + 2\sqrt{(x-2)(4-x)} = 6-x$

$\Rightarrow \qquad 2\sqrt{(x-2(4-x)} = 6-x-2$

$\Rightarrow \qquad 2\sqrt{(x-2)(4-x)} = 4-x$

Squaring again on both sides

$4(x-2)(4-x) = (4-x)^2$

$\Rightarrow \qquad (4-x)[4x-8-4+x] = 0$

$\Rightarrow \qquad (4-x)(5x-12) = 0$

$\Rightarrow \qquad x = 4, \; x = \dfrac{12}{5}$

Substitute $x = 4$, in (i)

L.H.S = $\sqrt{4-2} + \sqrt{4-4} = \sqrt{2}$

R.H.S = $\sqrt{6-4} = \sqrt{2}$

$\therefore \; x = 4$ is a solution.

Substitute $x = \dfrac{12}{5}$ in (i)

$$\text{L.H.S} = \sqrt{\frac{12}{5} - 2} + \sqrt{4 - \frac{12}{5}}$$

$$= \sqrt{\frac{2}{5}} + \sqrt{\frac{8}{5}} = \sqrt{\frac{2}{5}} + 2\sqrt{\frac{2}{5}} = 3\sqrt{\frac{2}{5}}$$

$$\text{R.H.S} = \sqrt{6 - \frac{12}{5}} = \sqrt{\frac{18}{5}} = 3\sqrt{\frac{2}{5}}$$

$$\therefore \quad x = \frac{12}{5} \text{ is also a solution.}$$

Note: Whenever we square a equation and find the roots, verify whether the roots satisfy initial equation or not.

16. $\log_{x^2+6x+8} \log_{2x^2+2x=3}(x^2 - 2x) = 0$

$\Rightarrow \qquad \log_{2x^2+2x+3}(x^2 - 2x) = (x^2 + 6x + 8)^0$

$\Rightarrow \qquad \log_{2x^2+2x+3}(x^2 - 2x) = 1$

$\Rightarrow \qquad x^2 - 2x = (2x^2 + 2x + 3)^1$

$\Rightarrow \qquad x^2 + 4x + 3 = 0$

$\Rightarrow \qquad (x+1)(x+3) = 0$

$\Rightarrow \qquad x = -1 \quad or \quad x = -3$

$x = -1$ *and* $x = -3$ satisfy the condition $x^2 - 2x > 0$

But at $x = -3$, $x^2 + 6x + 8 = 9 + 6(-3) + 8 = -1 < 0$ which is not possible

$\therefore \qquad x = -3$ is not the solution.

$x = -1$ satisfies $x^2 + 6x + 8 > 0$ and $2x^2 + 2x + 3 > 0$

Also at $x = -1$, $\qquad x^2 + 6x + 8 = 3 \neq 1 \qquad\qquad$ and

$$2x^2 + 2x + 3 \neq 1$$

Hence $x = -1$ is the only solution.

17. $\log_{2x+3}(6x^2 + 23x + 21) + \log_{3x+7}(4x^2 + 12x + 9) = 4$

$\Rightarrow \log_{(2x+3)}(2x+3)(3x+7) + \log_{(3x+7)}(2x+3)^2 = 4$

55

$$\log_{(2x+3)}(2x+3) + \log_{(2x+3)}(3x+7) + 2\log_{(3x+7)}(2x+3) = 4$$

$$1 + \log_{(2x+3)}(3x+7) + 2\log_{(3x+7)}(2x+3) = 4$$

Let $\log_{(2x+3)}(3x+7) = a$

then $\log_{(3x+7)}(2x+3) = \dfrac{1}{a}$

$$\Rightarrow \qquad a + \frac{2}{a} = 3$$

$$\Rightarrow \qquad a^2 - 3a + 2 = 0$$

$$\Rightarrow \qquad (a-1)(a-2) = 0$$

$$\Rightarrow \qquad a = 1 \quad or \quad a = 2$$

Consider: $a = 1$

$$\log_{(2x+3)}(3x+7) = 1$$

$$\Rightarrow \qquad 3x + 7 = 2x + 3$$

$$\Rightarrow \qquad x = -4$$

But $x = -4$ does not satisfy $2x+3 > 0$ and $3x+7 > 0$.

Hence $x = -4$ is not a solution.

Case (ii) $a = 2$

$$\log_{(2x+3)}(3x+7) = 2$$

$$\Rightarrow \qquad (3x+7) = (2x+3)^2$$

$$\Rightarrow \qquad 4x^2 + 9x + 2 = 0$$

$$\Rightarrow \qquad (4x+1)(x+2) = 0$$

$$\Rightarrow \qquad x = -\frac{1}{4} \quad or \quad x = -2$$

$x = -2$ does not satisfy $2x + 3 > 0$

Hence $x = -2$ is not a solution.

$x = -\dfrac{1}{4}$ satisfies $3x + 7 > 0$ and $2x + 3 > 0$

Also at $x = -\dfrac{1}{4}$, $2x + 3 \neq 1$

$\therefore \quad x = -\dfrac{1}{4}$ is the only solution.

18. Put $\dfrac{x-a}{x-b} = t$

$t + \dfrac{1}{t} = \dfrac{a}{b} + \dfrac{b}{a}$

$\Rightarrow \quad t^2 - t\left(\dfrac{a}{b} + \dfrac{b}{a}\right) + 1 = 0$

$\therefore \quad t = \dfrac{\dfrac{a}{b} + \dfrac{b}{a} \pm \sqrt{\left(\dfrac{a}{b} + \dfrac{b}{a}\right)^2 - 4}}{2}$

$= \dfrac{\dfrac{a}{b} + \dfrac{b}{a} \pm \left(\dfrac{a}{b} - \dfrac{b}{a}\right)}{2}$

$\Rightarrow \quad t = \dfrac{a}{b} \quad or \quad \dfrac{b}{a}$

$\therefore$ Case (i) $\quad t = \dfrac{a}{b}$

$\dfrac{x-a}{x-b} = \dfrac{a}{b}$

$\Rightarrow bx - ab = ax - ab \Rightarrow x(a-b) = 0 \Rightarrow x = 0$

Case (iii) $t = \dfrac{b}{a}$;

$\dfrac{x-a}{x-b} = \dfrac{b}{a}$

$\Rightarrow \quad ax - a^2 = bx - b^2$

$\Rightarrow \quad x(a-b) = a^2 - b^2$

$\Rightarrow \qquad x = a + b$

$\therefore \qquad x = 0, \; a + b$

19. Given equation can be written as

$$(x+1)(x^2 + 3x + 1) = 0$$

$\Rightarrow \qquad x = -1$

or $\qquad x^2 + 3x + 1 = 0$

$$x = \frac{-3 \pm \sqrt{9 - 4}}{2}$$

$\therefore \qquad x = -1, \; -\dfrac{3+\sqrt{5}}{2}, \; \dfrac{-3-\sqrt{5}}{2}$

20. The given equation can be written as

$$\left[x^2 + \frac{1+\sqrt{13}}{2} x + 1 \right]\left[x^2 + \frac{1-\sqrt{13}}{2} x + 1 \right] = 0$$

$\Rightarrow \qquad x^2 + \dfrac{1+\sqrt{13}}{2} x + 1 = 0 \qquad \qquad \text{... (i)}$

$\qquad \qquad x^2 + \dfrac{1-\sqrt{13}}{2} x + 1 = 0 \qquad \qquad \text{... (ii)}$

The first equation has two roots

$$x = \frac{-\sqrt{13} - 1 + \sqrt{2\sqrt{13} - 2}}{4}, \; \frac{-\sqrt{13} - 1 - \sqrt{2\sqrt{13} - 2}}{4}$$

Second equation has no real roots.

21. Let $x^2 = t$

$\therefore \qquad \quad t^2 - t - 6 = 0$

$\Rightarrow \qquad (t+2)(t-3) = 0$

$\Rightarrow \qquad t = -2 \quad or \qquad t = 3$

Case (i): $t = -2, \quad x^2 = -2$

$\therefore \qquad \qquad$ No real roots

Case (ii): $t = 3, \qquad x^2 = 3$

$$\therefore \qquad x = \pm\sqrt{3}$$

22. $M = b^2 - 4a(c - 2a)$

$$= (-10)^2 - 4(1)(26 - 2) = 100 - 96 = 4 > 0$$

$\therefore$ The equivalent equation is

$$\left[x^2 + \frac{(-10) + \sqrt{100 - 4(26 - 2)}}{2} x + 1 \right]$$

$$\left[x^2 + \frac{(-10) - \sqrt{100 - 4(26 - 2)}}{2} x + 1 \right] = 0$$

$$\Rightarrow \qquad \left(x^2 + \frac{-10 + 2}{2} x + 1 \right)\left(x^2 + \frac{-10 - 2}{2} x + 1 \right) = 0$$

$$\Rightarrow \qquad x^2 - 4x + 1 = 0 \qquad or \qquad x^2 - 6x + 1 = 0$$

$$\Rightarrow \qquad x = \frac{4 \pm \sqrt{16 - 4}}{2} \qquad or \qquad x = \frac{6 \pm \sqrt{36 - 4}}{2}$$

$$\Rightarrow \qquad x = 2 \pm \sqrt{3} \qquad or \qquad x = 3 \pm 2\sqrt{2}$$

23. $(2^2)^x - 3.2^x + 2 = 0$

$$\Rightarrow \qquad (2^x)^2 - 3.2^x + 2 = 0$$

Let $\qquad t = 2^x$

$$\therefore \qquad t^2 - 3t + 2 = 0$$

$$(t - 2)(t - 1) = 0$$

$$\Rightarrow \qquad t = 2 \qquad or \qquad t = 1$$

$$\therefore \qquad 2^x = 2^1 \qquad or \qquad 2^x = 1 = 2^0$$

$$\Rightarrow \qquad x = 1 \qquad or \qquad x = 0$$

24. $x^2 - 4x + 3 = x - 3$

$$\Rightarrow \qquad x^2 - 5x + 6 = 0$$

$$\Rightarrow \qquad (x - 2)(x - 3) = 0$$

$$\Rightarrow \qquad x = 2 \qquad or \qquad x = 3$$

Either $x = 2$ or $x = 3$ is not satisfying the original equation.

$\therefore$ The original equation has no roots.

25.   Rewrite the equation in the form

$$\frac{(a-x)^{\frac{3}{2}} + (x-b)^{\frac{3}{2}}}{(a-x)^{\frac{1}{2}} + (x-b)^{\frac{1}{2}}} = a - b$$

wherefrom we have,

$$a - x - (a-x)^{\frac{1}{2}}(x-b)^{\frac{1}{2}} + x - b = a - b$$

or $\sqrt{(a-x)(x-b)} = 0$

Thus, the required solutions will be

$$x_1 = a, \; x_2 = b$$

26.   We have

$$\sqrt{4a+b-5x} + \sqrt{4b+a-5x} = 3\sqrt{a+b-2x}$$

Squaring both members of the equality and performing all the necessary transformations, we get

$$\sqrt{4a+b-5x} \cdot \sqrt{4b+a-5x} = 2(a+b-2x)$$

Squaring them once again, we find

$$(4a+b)(4b+a) - 5x(4a+b+4b+a) + 25x^2 =$$

$$= 4(a^2 + b^2 + 4x^2 + 2ab - 4ax - 4bx)$$

Hence, $x^2 - ax - bx + ab = 0$

and, consequently, $x_1 = a, \; x_2 = b$.

Substituting the found values into the original equation, we get

$$\sqrt{b-a} + 2\sqrt{b-a} - 3\sqrt{b-a} = 0$$

$$2\sqrt{a-b} + \sqrt{a-b} - 3\sqrt{a-b} = 0$$

Hence, if $a \neq b$, then the equation has two roots; a and b (strictly speaking, if the operations with complex numbers are regarded as unknown, then there will be only one root).

27.   Rewrite the given equation as

$$(1+\lambda)x^2 - (a+c+\lambda b + \lambda d)x + ac + \lambda bd = 0$$

Set up the discriminant of this equation $D(\lambda)$. We have

$$D(\lambda) = \lambda^2(b-d)^2 + 2\lambda(ab + ad + bc + dc - 2bd - 2ac) + (a-c)^2$$

Set up the discriminant of this equation $D(\lambda)$. We have

$$D(\lambda) = (a + c + \lambda b + \lambda d)^2 - 4(1 + l)(ac + \lambda bd) .$$

On transformation we obtain

$$D(\lambda) = \lambda^2(b-d)^2 + 2\lambda(ab + ad + bc + dc - 2bd - 2ac) + (a-c)^2$$

We have to prove that $D(\lambda) \geq 0$ for any λ. Since $D(\lambda)$ is a second-degree trinomial in λ and $D(0) = (a-c)^2 > 0$, it is sufficient to prove that the roots of this trinomial are imaginary. And for the roots of our trinomial to be imaginary, it is necessary and sufficient that the expression

$$4(ab + ad + bc + dc - 2bd - 2ac)^2 - 4(a-c)^2(b-d)^2$$

be less than zero. We have

$$4(ab + ad + bc + dc - 2bd - 2ac)^2 - 4(a-c)^2(b-d)^2 =$$

$$= 4(ab + ad + bc + dc - 2bd - 2ac - ab + cb + ad - cd) \times$$

$$(ab + ad + bc + dc - 2bd - 2ac + ab - cd - ad + cd) =$$

$$-16(b-a)(d-c)(c-b)(d-a)$$

The last expression is really less than zero by virtue of the given conditions

$$a < b < c < d$$

28. The original equation can be rewritten in the following way

$$3x^2 - 2(a + b + c)x + ab + ac + bc = 0$$

Let us the prove that

$$2(a + b + c)^2 - 12(ab + ac + bc) \geq 0$$

We have,

$$4(a + b + c)^2 - 12(ab + ac + bc) = 4(a^2 + b^2 + c^2 - ab - ac - bc)$$

$$= 2(2a^2 + 2b^2 + 2c^2 - 2ab - 2ac - 2bc)$$

$$= 2\left\{(a^2 - 2ab + b^2) + (a^2 - 2ac + c^2) + (b^2 - 2bc + c^2)\right\}$$

61

$$= 2\{(a-b)^2 + (a-c)^2 + (b-c)^2\} \geq 0$$

29. Suppose the roots of both equations are imaginary.

Then, $p^2 - 4q < 0,\ p_1^2 - 4q_1 < 0$

Consequently, $p^2 + p_1^2 - 4q - 4q_1 < 0,\ p^2 + p_1^2 - 2pp_1 < 0, (p - p_1)^2 < 0$

which is impossible.

30. Let us rewrite the given equation as

$$(a+b+c)x^2 - 2(ab+ac+bc)x + 3abc = 0$$

Prove that its discriminant is greater than or equal to zero.

We have, $4(ab+ac+bc)^2 - 12abc(a+b+c) =$

$$2\{(ab-ac)^2 + (ab-bc)^2 + (ac-bc)^2\} \geq 0$$

31. By properties of the quadratic equation we have the following system

$$p + q = -p,\ pq = q$$

From the second equation we get,

$$q(p-1) = 0$$

Hence, either $q = 0$ or $p = 1$. From the first one we find

if $q = 0$, then $p = 0$; if $p = 1$, then $q = -2$.

Thus, we have two quadratic equations satisfying the set requirements

$$x^2 = 0 \ and \ x^2 + x - 2 = 0$$

32. $\left(x + \dfrac{1}{x}\right)^2 - \dfrac{3}{2}\left(x - \dfrac{1}{x}\right) = 4$

$\Rightarrow \left(x - \dfrac{1}{x}\right)^2 + 4 - \dfrac{3}{2}\left(x - \dfrac{1}{x}\right) = 4 \Rightarrow \left(x - \dfrac{1}{x}\right)^2 - \dfrac{3}{2}\left(x - \dfrac{1}{x}\right) = 0$

$\Rightarrow t^2 - \dfrac{3}{2}t = 0 \ where \ t = x - \dfrac{1}{x}$

$\Rightarrow t\left(t - \dfrac{3}{2}\right) = 0 \Rightarrow t = 0 \ or \ t = 3/2$

$$\Rightarrow x - \frac{1}{x} = 0 \ or \ x - \frac{1}{x} = \frac{3}{2}$$

$$\Rightarrow x^2 = 1 \ or \ 2x^2 - 3x - 2$$

$$\Rightarrow x = \pm 1 \ or \ 2x^2 - 4x + x - 2 = 0$$

$$\Rightarrow x = \pm 1 \ or \ (x-2)(2x+1) = 0$$

$$\Rightarrow x = \pm 1 \ or \ 2 \ or \ -1/2$$

$$\Rightarrow Solution \ set = \{\pm 1, 2, -1/2\}$$

33. $x^4 - 2x^3 - x^2 - 2x + 1 = 0$

$$\Rightarrow x^2 - 2x - 1 - \frac{2}{x} + \frac{1}{x^2} = 0$$

$$\Rightarrow \left(x^2 + \frac{1}{x^2} \right) - 2\left(x + \frac{1}{x} \right) - 1 = 0$$

$$\Rightarrow \left(x + \frac{1}{x} \right)^2 - 2 - 2\left(x + \frac{1}{x} \right) - 1 = 0$$

$$\Rightarrow y^2 - 2y - 3 = where \ y = x + \frac{1}{x}$$

$$\Rightarrow (y-3)(y+1) = 0$$

$$\Rightarrow y - 3 = 0 \qquad or \qquad y + 1 = 0$$

$$\Rightarrow x + \frac{1}{x} - 3 = 0 \ or \ x + \frac{1}{x} + 1 = 0$$

$$\Rightarrow x^2 - 3x + 1 = 0 \ or \ x^2 + x + 1 = 0$$

$$\Rightarrow x = \frac{3 \pm \sqrt{9-4}}{2} \ or \ x = \frac{-1 \pm \sqrt{1-4}}{2}$$

$$\Rightarrow x = \frac{3 \pm \sqrt{5}}{2} \ or \ \frac{-1 \pm \sqrt{3}i}{2}$$

34. $(x-1)(x-3)(x-5)(x-7) = 9$

$$\Rightarrow \left[(x-7)(x-1) \right] \left[(x-3)(x-5) \right] = 9$$

63

$\Rightarrow (x^2 - 8x + 7)(x^2 - 8x + 15) = 9$

$\Rightarrow (t + 7)(t + 15) = 9 \ where \ t = x^2 - 8x$

$\Rightarrow t^2 + 22t + 105 - 9 = 0$

$\Rightarrow t^2 + 22t + 96 = 0 \Rightarrow t^2 + 16t + 6t + 96 = 0$

$\Rightarrow t(t + 16) + 6(t + 16) = 0 \Rightarrow (t + 16)(t + 6) = 0$

$\Rightarrow (x^2 - 8x + 16)(x^2 - 8x + 6) = 0$

$\Rightarrow x^2 - 8x + 16 = 0 \ or \ x^2 - 8x + 6 = 0$

$\Rightarrow (x - 4)^2 = 0 \ or \ x = \dfrac{8 \pm \sqrt{64 - 24}}{2}$

$\Rightarrow x = 4, 4 \ or \ x = 4 \pm \sqrt{10}$

$\Rightarrow Solution \ set = \left\{ 4, \ 4 \pm \sqrt{10} \right\}$

35.   Given equation is $kx^2 - 14x + 8 = 0$

      Let the roots be $\alpha, 6\alpha$

      Sum of the roots, $\alpha + 6\alpha = 14/k \Rightarrow 7\alpha = 14/k \Rightarrow \alpha = 2/k$
      Product of the roots,

      $\alpha(6\alpha) = 8/k \Rightarrow 3\alpha^2 = 4/k \Rightarrow 3(2/k)^2 = 4/k \Rightarrow k = 3$

36.   Let $f(x) = x^2 + x - 1$

      Equation whose roots are $\alpha + 2, \ \beta + 2$ is $f(x - 2) = 0$

      $f(x - 2) = (x - 2)^2 + (x - 2) - 1 = 0$

      $\Rightarrow x^2 - 3x + 1 = 0$

37.   If one root of the equation is $2 + \sqrt{3}$ then other root is $2 - \sqrt{3}$

      Sum of the roots = $2 + \sqrt{3} + 2 - \sqrt{3} = 4 = -p \Rightarrow p = -4$

      Product of the roots = $(2 + \sqrt{3})(2 - \sqrt{3}) = q \Rightarrow q = 4 - 3 = 1$

38.   Squaring both sides

      $2x - 6 + x + 4 + 2\sqrt{(2x - 6)(x + 4)} = 25$

$$\Rightarrow 2\sqrt{(2x-6)(x+4)} = 25-4+6-3x = 27-3x$$

Again squaring both sides and simplifying.

We arrive at the equation $x^2 - 170x + 825 = 0 \Rightarrow x = 5, 165$

Direct substitution of these values in the original equation shows that $x = 5$ is root and $x = 165$ is not.

Why $x = 165$ is not a root of an equation?

Reason: *(Extraneous root appeared due to squaring the equation)*

39. Any quadratic equation $ax^2 + bx + c = 0$ $(a \neq 0)$ has infinite solutions then $a = 0, b = 0, c = 0$.

$$\left(\because \quad 0.x^2 + 0.x + 0 = 0 \text{ is satisfied for all real } 'x' \right)$$

$$\therefore \quad \left(a^2 - 4a + 3\right) x^2 + (a-1)x + a^2 - 1 = 0 \text{ has infinite roots if}$$

$$(a^2 - 4a + 3) = 0, \ (a-1) = 0, \ a^2 - 1 = 0 \text{ simultaneously.}$$

that is possible only for $a = 1$.

40. $px^2 + qx + r = 0$ have real roots

$$\Rightarrow D = 4q^2 - 2pr \geq 0 \Rightarrow q^2 \not< pr \qquad \qquad \dots (1)$$

$$qx^2 - 2\sqrt{pr}x + q = 0 \text{ have real roots}$$

$$\Rightarrow D = 4qr - 4q^2 \geq 0 \Rightarrow pr - q^2 \geq 0$$

$$\Rightarrow pr \geq q^2 \Rightarrow q^2 \leq pr \Rightarrow q^2 \not> pr \qquad \qquad \dots (2)$$

From (1) and (2) $q^2 \not< pr$ and $q^2 \not> pr \Rightarrow q^2 = pr$

41. Let $S = (1 + a + a^2 + \dots \dots \infty)(1 + b + b^2 + \dots \dots \infty)$

$$S_\infty = \frac{a}{1-r} \cdot S = \left(\frac{1}{1-a}\right)\left(\frac{1}{a-b}\right)$$

$$S = \frac{1}{1-a-b+ab}, \ S = \frac{1}{1-(a+b)+ab}$$

a, b are the roots of $11x^2 - 4x - 2 = 0 \Rightarrow a+b = 4/11, \ ab = -2/11$

$$S = \cfrac{1}{1 - \cfrac{4}{11} - \cfrac{2}{11}} \Rightarrow S = \cfrac{1}{1 - \cfrac{6}{11}} = \frac{11}{5}$$

42. Consider example $x^2 + 8x + 7 = 0$, $ax^2 + bx + c = 0$

$ac = 7$ which can't be expressed as sum of square of two natural numbers

$\therefore$ 'a' is wrong.

Consider same example $x^2 - 8x + 7 = 0$ where a, b, c are integers and roots also natural numbers '1' and '7'.

$a = 1$, $b = -8$, $c = 7$ 'b' does not divide 'c'.

Hence 'c' is wrong

'c' does not divide 'b' in the same example

Hence 'd' is wrong.

Let α, β be the roots of $ax^2 + bx + c = 0$

$$\Rightarrow \alpha + \beta = \frac{-b}{a}, \ \alpha\beta = \frac{c}{a}$$

α, β are integers $\Rightarrow \alpha + \beta$, $\alpha\beta$ are integers

$$\Rightarrow \frac{-b}{a}, \frac{c}{a} \text{ are integers} \Rightarrow \text{'a' divide 'b' \& 'c'}$$

$\Rightarrow$ 'b' is correct

43. $(1 + a^2)x^2 + 2(x - a)(1 + ax) + 1$

$$\Rightarrow (1 + a^2)x^2 + 2(x + ax^2 - a - a^2x) + 1$$

$$\Rightarrow (1 + a^2)x^2 + x(2 - 2a^2) + 2ax^2 + 1 - 2a$$

$$\Rightarrow (1 + a^2 + 2a)x^2 + (2 - 2a^2)x + (1 - 2a)$$

$$\Rightarrow (1 + a)^2 x^2 + 2(1 - a^2)x + (1 - 2a)$$

$$D = B^2 - 4AC$$

$$D = 4(1 - a^2)^2 - 4(1 + a)^2(1 - 2a)$$

$$D = (1 + a^2 4\left((1 - a)^2 - (1 - 2a)\right)$$

$$D = 4(1+a)^2 (1+a^2 - 2a - 1 + 2a)$$

$$D = 4a^2 (a+1)^2 \geq 0$$

$\therefore$ For all value of 'a' roots are real.

44. α, β are roots of

$$x^2 + ax + b = 0 \Rightarrow \alpha + \beta = -a, \ \alpha\beta = b$$

$$\Rightarrow \frac{\alpha + \beta}{\alpha\beta} = \frac{-a}{b} \Rightarrow \frac{1}{\beta} + \frac{1}{\alpha} = -\frac{a}{b}$$

γ, δ are roots of $x^2 - ax + (b-2) = 0 \Rightarrow \gamma + \delta = a, \gamma\delta = b-2$

$$\Rightarrow \frac{\gamma + \delta}{\gamma\delta} = \frac{a}{b-2} \Rightarrow \frac{1}{\gamma} + \frac{1}{\delta} = \frac{a}{b-2}$$

$$\alpha\beta\gamma\delta = 24 \Rightarrow b(b-2) = 24$$

$$\Rightarrow b^2 - 2b - 24 = 0 \Rightarrow b^2 - 6b + 4b - 24 = 0$$

$$\Rightarrow b(b-6) + 4(b-6) = 0 \Rightarrow (b+4)(b-6) = 0$$

$$\Rightarrow b = -4, 6$$

$$\Rightarrow \frac{1}{\alpha} + \frac{1}{\beta} + \frac{1}{\gamma} + \frac{1}{\delta} = \frac{-a}{b} + \frac{a}{b-2} = \frac{5}{12}$$

If $b = -4 \Rightarrow \dfrac{-a}{-4} + \dfrac{a}{-6} = \dfrac{5}{12}$

$$\Rightarrow \frac{a}{4} - \frac{a}{6} = \frac{5}{12} \Rightarrow \frac{3a - 2a}{12} = \frac{5}{12} \Rightarrow \frac{a}{12} = \frac{5}{12} \Rightarrow a = 5$$

45.   The equation can be written as

$$3x^2 - 4x + 4 - 4\sqrt{3x^2 - 4x + 1} = 0$$

$\Rightarrow$ $\quad 3x^2 - 4x + 1 - 4\sqrt{3x^2 - 4x + 1} + 3 = 0$

Put $\quad 3x^2 - 4x + 1 = t^2$

$\Rightarrow$ $\quad t^2 - 4t + 3 = 0$

$\Rightarrow$ $\quad (t-1)(t-3) = 0$

$\Rightarrow$ $\quad t = 1 \qquad or \qquad t = 3$

$\therefore \qquad \sqrt{3x^2 - 4x + 1} = 1 \quad or \ 3$

$\Rightarrow \qquad 3x^2 - 4x = 0 \qquad or \qquad 3x^2 - 4x - 8 = 0$

$\Rightarrow \qquad x = \dfrac{4}{3}, 0, \dfrac{2}{3}\left(1 \pm \sqrt{6}\right)$

EXERCISE - 2   [Time: 3hrs]

1. Solve $ab(x^2 + 1) = (a^2 + b^2)x$

2. Solve: $16\left(x^2 + \dfrac{1}{x^2}\right) - 257 = 0$

3. Solve: $\dfrac{2a\sqrt{1 + x^2}}{x + \sqrt{1 + x^2}} = a + b$

4. Solve: $3^{x-2} + 3^{3-x} = 4$

5. Solve: $(x + 2)(x + 3)(x + 4)(x + 5) = 24(x^2 + 7x + 7)$

6. Solve: $\sqrt{x^2 - 3x + 36} - \sqrt{x^2 - 3x + 9} = 3$

7. Solve: $\sqrt{x + 5} + \sqrt{x + 12} = \sqrt{2x + 41}$

8. Solve: $3x^4 - 20x^3 - 94x^2 - 20x + 3 = 0$

9. $\sqrt{x^2 + 4x - 21} + \sqrt{x^2 - x - 6} = \sqrt{6x^2 - 5x - 39}$ (factorise)

EXERCISE - 2
Answers

1. $a/b \ (or) \ b/a$

2. $4 \ (or) \ \dfrac{1}{4} \ (or) -4 \ (or) -\dfrac{1}{4}$

3. 4

4. $2 \ (or) \ 3$

5. $\{-1, 1, -6, -8\}$

6. $0 \ (or) \ 3$

7. 4

8. $-\dfrac{1}{3}, -3, 5 + \sqrt{24}, 5 - \sqrt{24}$

4. SYSTEMS OF ALGEBRAIC EQUATIONS

1. A linear equation in n unknowns is an equation of the form

$$a_1 x_1 + a_2 x_2 + \ldots\ldots a_n x_n = b$$

where $a_1, a_2, \ldots\ldots, a_n, b$ are numbers

2. Consider equations

$$x + 2y + 3z = 8$$

$$3x + y^2 + z = 6$$

$$2x^2 + y + 2z = 6$$

If the equations considered simultaneously constitute a system of equations.

A solution of the system consists the values of the unknown.

A system of equations which has solution is said to be consistent, and that which has no solutions is said to be in consistent.

A system of equations is linear if all equations are linear.

3. Equations in two unknowns

$$a_0 x^n + a_1 x^{n-1} y + a_2 x^{n-2} y^2 + \ldots + a_r x^{n-r} y^r + \ldots + a_n y^n = 0$$

are called homogeneous equations.

In a homogeneous equation each term contains a product of the powers of x and y the sum of whose exponents is constant.

4. A system of equations in n unknowns $x_1, x_2, \ldots\ldots, x_n$ is said to be symmetric if it does not change upon a permutation of the unknowns.

If there are two unknowns $(x \text{ and } y)$ then the solution of such systems can be found by introducing new unknowns

$$u = x + y \text{ and } v = xy$$

Symmetric systems of three equations in three unknowns x, y and z are usually solved by introducing new unknowns.

$$u = x + y + z$$

$$v = xy + yz + zx$$

$$w = xyz$$

EXERCISE - 1

1. Solve the system of equations

    $$x + 2y + 3z = 8$$

    $$3x + y + z = 6$$

    $$2x + y + 2z = 6$$

2. Solve the system of equations

    $$a_1 x + b_1 y + c_1 = 0$$

    $$a_2 x + b_2 y + c_2 = 0$$

3. Solve the system of equations?

    $$x^2 - 5xy + 6y^2 = 0$$

    $$x^2 + y^2 = 10$$

4. Solve the system of equations

    $$x^2 + y^2 = 2(xy + 2)$$

    $$x + y = 6$$

5. Solve the system of equations

    $$\frac{3xy}{x+y} = 5, \qquad \frac{2xz}{x+z} = 3, \qquad \frac{yz}{y+z} = 4$$

6. Solve the system of equations

    $$xy + x + y = 11$$

    $$x^2 y + xy^2 = 30$$

7. Solve: $x^2 + y^2 + 6x + 2y = 0$

    $$x + y + 8 = 0$$

8. Solve: $\dfrac{x}{y} + \dfrac{y}{x} = \dfrac{13}{6}$

    $$x + y = 5$$

9. Solve $\dfrac{x+y}{x-y} + 6\,\dfrac{x-y}{x+y} = 5$; $xy = 2$

10. Solve: $x^2 + xy = 15$

$y^2 + xy = 10$

11. Solve $x + y + xy = 5$

$x^2 + y^2 + xy = 7$

12. Solve in real numbers the system of equations.

$x + y + z = 4$

$x^2 + y^2 + z^2 = 14$

$x^3 + y^3 + z^3 = 34$

13. Solve in real numbers the system of equations:

$a + b = 8$

$ab + c + d = 23$

$ad + bc = 28$

$cd = 12$

14. Solve:

$(x + y)(x + z) = 15$

$(y + z)y + x) = 18,$

$(z + x)(z + y) = 30,$

15. Solve: $x^2 + xy + xz = -12$

$y^2 + yz + yz = 30$

$z^2 + zx + zy = 18$

16. Solve: $y^2 + yz + z^2 = 19,$ ... (1)

$z^2 + zx + x^2 = 39,$ ... (2)

$x^2 + xy + y^2 = 49$ ... (3)

17. Solve: $x^3 + y^3 + z^3 = a^3, x^2 + y^2 + z^2 = a^2, x + y + z = a,$

18. If $a + b + c = 1, a^2 + b^2 + c^2 = 9$ and $a^3 + b^3 + c^3 = 1,$ find

71

$$\frac{1}{a} + \frac{1}{b} + \frac{1}{c}$$

19. Solve: $3x(x + y - 2) = 2y$,

    $y(x + y - 1) = 9x$

20. Solve: $\log_3(\log_2 x) + \log_{1/3}(\log_{1/2} y) = 1$, $xy^2 = 4$

21. Solve:

    $$\log_2 x + \log_4 y + \log_4 z = 2$$

    $$\log_3 y + \log_9 z + \log_9 x = 2$$

    $$\log_4 z + \log_{16} x + \log_{16} y = 2$$

22. Solve:

    $$(x + y)(x + y + z) = 18 ,$$

    $$(y + z)(x + y + z) = 30 ,$$

    $$(z + x)(x + y + z) = 2A ,$$

23. Solve the equations:

    $$(x - 4)(y - 4) = 16 ,$$

    $$(y - 6)(z - 6) = 36,$$

    $$(z - 8)(x - 8) = 64 ,$$

24. Solve completely the following system of equations:

    $$(x - 2)(y - 2) = 4,$$

    $$(y - 3)(z - 3) = 9,$$

    $$(z - 4)(x - 4) = 16,$$

25. Solve the system of equations

    $$xy + x + y = -13,$$

    $$yz + y + z = -9,$$

    $$zx + z + x = 5,$$

26. Solve the system of equations for real x and y :

$$5x\left(1+\frac{1}{x^2+y^2}\right)=12; \ 5y\left(1-\frac{1}{x^2+y^2}\right)=4$$

27. Solve the following system of equations for real x, y, z :

$$x+y-z=4$$

$$x^2-y^2+z^2=-4$$

$$xyz=6 \ .$$

EXERCISE - 1
Solutions

1. $-3\times\left[(x+2y+3z)=8\right]$

   $\Rightarrow$ $-3x-6y-9z=-24$

        $3x+y+z=6$

Adding these two equations, we get

        $5y+8z=18$

        $-2\times\left[(x+2y+3z)=8\right]$

        $2x+y+2z=6$

Adding these two equations, we get

        $3y+4z=10$

        $3\times\left[5y+8z=18\right]$

        $-5\times\left[3y+4z=10\right]$

Adding these two equations, we get

        $4z=4$

   $\Rightarrow$ $z=1$

Given system is reduced to equivalent form

        $x+2y+3z=8$

        $5y+8z=18$

Substituting $z=1$ in 2nd equation, we get $y=2$

Substituting $z = 1$ and $y = 2$ in first equation, $x = 1$

$\therefore$　　　　$x = 1, \ y = 2, \ z = 1$

2.　　This can be solved by using the formula:

$$\frac{x}{b_1 c_2 - b_2 c_1} = \frac{-y}{a_1 c_2 - a_2 c_1} = \frac{1}{a_1 b_2 - a_2 b_1}$$

3.　　Consider $x^2 - 5xy + 6y^2 = 0$

Divide with y^2

$\Rightarrow$　　　$\left(\dfrac{x}{y}\right)^2 - 5\left(\dfrac{x}{y}\right) + 6 = 0$

Put　　　$\dfrac{x}{y} = t$

$\therefore$　　　$t^2 - 5t + 6 = 0$

$\Rightarrow$　　　$(t - 3)(t - 2) = 0$

$\Rightarrow$　　　$t = 3 \ \ or \ t = 2$

$\Rightarrow$　　　$x = 2y \quad or \quad x = 3y$

Substituting these values in second equation,

　　　　　$4y^2 + y^2 = 10 \quad \Rightarrow \quad y^2 = 2$

and　　$9y^2 + y^2 = 10 \quad \Rightarrow \quad y^2 = 1$

$\Rightarrow$　　$y = \pm 1 \ \ and \ y = \pm\sqrt{2}$

$\therefore$　　　$x = \pm 3 \quad and \quad x = \pm 2\sqrt{2}$

$\therefore$　　$(3, 1), \ (-3, -1), \left(2\sqrt{2}, \ \sqrt{2}\right), \left(-2\sqrt{2}, \ -\sqrt{2}\right)$

4.　　Put $u = x + y$ and $v = xy$

$\therefore \ x^2 + y^2 = (x + y)^2 - 2xy = u^2 - 2v$

$\therefore$ from first and second equations respectively

　　　　　$u^2 - 2v = 2v + 4$

　　　　　　　$u = 6$

$$\therefore \quad u^2 = 4v + 4$$

$$\Rightarrow \quad 36 = 4v + 4$$

$$\Rightarrow \quad v = 8$$

$$\therefore \quad x + y = 6$$

$$xy = 8$$

$$(x - y)^2 = (x + y)^2 - 4xy$$

$$(x - y)^2 = 36 - 32 = 4$$

$$\Rightarrow \quad x - y = \pm 2$$

$$\therefore \quad x = 4 \quad and \quad y = 2$$

$$\text{or} \quad x = 2 \quad and \quad y = 4$$

5. $\dfrac{x+y}{xy} = \dfrac{3}{5}, \dfrac{x+z}{xz} = \dfrac{2}{3}, \dfrac{y+z}{yz} = \dfrac{1}{4}$

$$\Rightarrow \quad \frac{1}{y} + \frac{1}{x} = \frac{3}{5}, \qquad \frac{1}{x} + \frac{1}{z} = \frac{2}{3}, \qquad \frac{1}{y} + \frac{1}{z} = \frac{1}{4}$$

Put $\quad \dfrac{1}{x} = u, \quad \dfrac{1}{y} = v, \quad \dfrac{1}{z} = w$

$$\therefore \quad u + v = \frac{3}{5}, u + w = \frac{2}{3}, w + v = \frac{1}{4}$$

Solving, we get $u = \dfrac{61}{120}, v = \dfrac{11}{120}, w = \dfrac{19}{120}$

$$\therefore \quad x = \frac{120}{61}, \qquad y = \frac{120}{11}, \qquad z = \frac{120}{19}$$

6. Let $u = x + y$ and $v = xy$

$$u + v = 11$$

$$x^2 y + xy^2 = 30 \quad \Rightarrow \quad xy(x + y) = 30$$

$$\Rightarrow \quad uv = 30$$

$$\therefore \quad u - v = \pm \sqrt{(u + v)^2 - 4uv}$$

75

$$= \pm\sqrt{121 - 4(30)} = \pm 1$$

$$\therefore \quad \left. \begin{array}{l} u + v = 11 \\ u - v = 1 \end{array} \right\} \quad \dots (i)$$

$$\left. \begin{array}{l} u + v = 11 \\ u - v = -1 \end{array} \right\} \quad \dots (ii)$$

Solving (i), we get

$$u = 6, \, v = 5$$

$$\therefore \quad x + y = 6$$

$$xy = 5$$

$$x - y = \pm\sqrt{6^2 - 4(5)} = \pm\sqrt{16} = \pm 4$$

$$\therefore \quad \left. \begin{array}{l} x + y = 6 \\ x - y = 4 \end{array} \right\} \quad \dots (a)$$

$$\begin{array}{l} x + y = 6 \\ x - y = -4 \end{array} \quad \dots (b)$$

from (a), we get $x = 5, \, y = 1$

from (b), we get $x = 1 \,\, y = 5$

Solving (ii), we get, $u = 5, \, v = 6$

$$\therefore \quad x + y = 5; \,\, xy = 6; \,\, x - y = \pm 1$$

Solving we get $x = 3, \, y = 2$ and $x = 2, \, y = 3$

Solution: $(5, 1)$, $(1, 5)$, $(3, 2)$ *and* $(2, 3)$

7. From second equation $y = -8 - x$

$$\therefore \quad (x)^2 + (-8 - x)^2 + 6x + 2(-8 - x) = 0$$

$$\Rightarrow \quad x^2 + 64 + x^2 + 16x + 6x - 16 - 2x = 0$$

$$\Rightarrow \quad 2x^2 + 20x + 48 = 0$$

$$\Rightarrow \quad x^2 + 10x + 24 = 0$$

$$(x + 6)(x + 4) = 0$$

$\Rightarrow \qquad x = -6,\ x = -4$

$\therefore \qquad y = -8 + 6 = -2$

$\text{and} \qquad y = -8 + 4 = -4$

$\therefore \qquad$ Solution: $(-6,\ -2),\ (-4,\ -4)$

8. Put $\dfrac{x}{y} = t$ in first equation, then

$$\therefore \qquad t + \frac{1}{t} = \frac{13}{6}$$

$$\Rightarrow \qquad (t^2 + 1)6 = 13t$$

$$\Rightarrow \qquad 6t^2 - 13t + 6 = 0$$

$$t = \frac{13 \pm \sqrt{169 - 4 \times 36}}{12} = \frac{13 \pm \sqrt{25}}{12}$$

$$\Rightarrow \qquad t = \frac{13 \pm 5}{2} \qquad \Rightarrow \qquad t = \frac{3}{2}\ or\ t = \frac{2}{3}$$

$$\therefore \qquad \frac{x}{y} = \frac{3}{2} \qquad or \qquad \frac{x}{y} = \frac{2}{3}$$

$$\Rightarrow \qquad 2x = 3y \qquad or \qquad 3x = 2y$$

Now from second equation

$$x + y = 5$$

$$\frac{3}{2}y + y = 5 \qquad or \qquad \frac{2}{3}y + y = 5$$

$$\Rightarrow \qquad 5y = 10 \qquad or \qquad 5y = 15$$

$$\Rightarrow \qquad y = 2 \qquad or \qquad y = 3$$

$$\therefore \qquad x = 3 \qquad or \qquad x = 2$$

$\therefore$ Solution: $(3,\ 2),\quad (2,\ 3)$

9. Put $\dfrac{x+y}{x-y} = t$ in first equation, then

$$\therefore \qquad t + \frac{6}{t} = 5$$

$$\Rightarrow \qquad t^2 + 6 = 5t$$

$$\Rightarrow \qquad t^2 - 5t + 6 = 0$$

$$(t = 2)\,(t = 3) = 0$$

$$\Rightarrow \qquad t = 2 \quad or \quad t = 3$$

$$\therefore \qquad \frac{x+y}{x-y} = 2 \quad or \quad \frac{x+y}{x-y} = 3$$

$$x + y = 2x - 2y \quad or \quad x + y = 3x - 3y$$

$$\Rightarrow \qquad x = 3y \quad or \quad 2x = 4y$$

$$\Rightarrow \qquad x = 3y \quad or \quad x = 2y$$

Substituting in second equation,

$$xy = 2$$

$$\Rightarrow \qquad 3y(y) = 2 \quad or \quad (2y)(y) = 2$$

$$\Rightarrow \qquad y^2 = \frac{2}{3} \quad or \quad y^2 = 1$$

$$\Rightarrow \qquad y = \pm\sqrt{\frac{2}{3}} \quad or \quad y = \pm 1$$

Solution is: $\left[3\sqrt{\frac{2}{3}}, \sqrt{\frac{2}{3}} \right], \left[-3\sqrt{\frac{2}{3}}, -\sqrt{\frac{2}{3}} \right]$

$$[2, +1], \ [-2, -1]$$

10. Adding two equations, we get

$$x^2 + 2xy + y^2 = 25$$

$$\Rightarrow \qquad (x + y)^2 = 25$$

$$\Rightarrow \qquad x + y = \pm 5$$

Using the second equation

$$y(x + y) = 10$$

$\Rightarrow \quad y(5) = 10 \qquad \Rightarrow \qquad y = 2, \quad x = 3$

$\quad\quad y(-5) = 10 \qquad \Rightarrow \qquad y = -2, \quad x = -3$

$\therefore$ Solution is $(3, 2), (-3, -2)$

11. Adding the two equations, we get

$$x^2 + y^2 + 2xy + x + y = 12$$

$\Rightarrow \quad (x + y)^2 + (x + y) - 12 = 0$

Put $\quad x + y = t$ in this equation

$$t^2 + t - 12 = 0$$

$$t^2 + 4t - 3t - 12 = 0$$

$$t(t + 4) - 3(t + 4) = 0$$

$\Rightarrow \quad (t + 4)(t - 3) = 0$

$$t = -4 \quad or \quad t = 3$$

from first equation

$$xy = 5 - (x + y) = 5 - t$$

$\Rightarrow \quad \left.\begin{array}{l} xy = 9 \\ x + y = -4 \end{array}\right\} \qquad \ldots \text{(a)}$

or $\quad \left.\begin{array}{l} xy = 2 \\ x + y = 3 \end{array}\right\} \qquad \ldots \text{(b)}$

Solving, we get from (b)

$$x = 1, \ y = 2 \qquad or \qquad x = 2, \ y = 1$$

$$x + y = -4$$

$$xy = 9$$

$$(x + y)^2 - 4xy = (x - y)^2$$

$\Rightarrow \quad (x - y)^2 = 16 - 36 = -20$

$$x - y = \pm\sqrt{20}\ i$$

No real solution for this case.

12. Let $p(u) = u^3 + au^2 + bu + c$ with roots x, y, z.

$$x + y + z = -a$$

$$xy + yz + zx = b$$

∴ from the given equations, $a = -4$

∵ $x^2 + y^2 + z^2 = (x + y + z)^2 - 2(xy + yz + zx)$

$$14 = (4)^2 - 2(xy + yz + zx)$$

$\Rightarrow$ $2(xy + yz + zx) = 2$

$\Rightarrow$ $xy + yz + zx = 1$

$\Rightarrow$ $b = 1$

∵   The numbers x, y, z are the roots of p(u).

∴ $x^3 - 4x^2 + x + c = 0$

$$y^3 - 4y^2 + y + c = 0$$

$$z^3 - 4z^2 + z + c = 0$$

Adding these equations, we get

$$(x^3 + y^3 + z^3) - 4(x^2 + y^2 + z^2) + (x + y + z) + 3c = 0$$

$\Rightarrow$ $34 - 4(14) + 4 + 3c = 0$

$\Rightarrow$ $3c = 18 \Rightarrow \quad c = 6$

∴ $P(u) = u^3 - 4u^2 + t + 6$

$u_1 = -1$ is a root of the equation

∴ $P(u) = (u + 1)(u^2 - 5u + 6)$

$\Rightarrow$ $u_2 = 2$ and $u_3 = 3$

∴ The Solution of the system are the triple $(-1, 2, 3)$ and all of its permutations.

13.   Consider the product of equations

$$(x^2 + ax + c)(x^2 + bx + d)$$

$=$ $x^4 + (a + b)x^3 + (ab + c + d)x^2 + (ad + bc)x + cd$

∴ $P(x) = x^4 + 8x^3 + 23x^2 + 28x + 12$

factorising $P(x)$, we get

$$P(x) = (x+1)(x+2)^2(x+3)$$

The polynomial factors in two ways as a product of quadratic polynomials.

$$P(x) = (x^2 + 4x + 3)(x^2 + 4x + 4)$$

and $\quad P(x) = (x^2 + 3x + 2)(x^2 + 5x + 6)$

$\therefore$ Solution:

(4, 4, 3, 4) , (4, 4, 4, 3)

(3, 5, 2, 6) , (5, 3, 6, 2)

14. Let $y + z = X$, $z + x = Y$, $x + y = Z$

then $YZ = 15$, $ZX = 18$, $XY = 30$ ... (i)

Multiplying the corresponding sides of these equations, we have

$$X^2 Y^2 Z^2 = 30 \times 18 \times 15 = (15 \times 6)^2$$

i.e., $\quad XYZ = \pm 90$

When $XYZ = 90$, we get using equations (1),

$$X = 6, Y = 5, Z = 3$$

i.e., $\quad y + z = 6$, $z + x = 5$, $x + y = 3$,

so that $\quad x = 1, y = 2, z = 4$.

When $XYZ = -90$, we get using equations (1),

$X = -6, Y = -5, Z = -3$

i.e., $y + 2 = -6$, $z + x = -5$, $x + y = -3$,

so that $\quad x = -1, y = -2, z = -4$.

Hence the solutions are

$x = 1, y = 2, z = 4 \qquad$ or $\qquad x = -1, y = -2, z = -4$

15. The given equations can be re-written as

$$x(x + y + z) = -12,$$

$$y(x + y + z) = 30,$$

$$z(x + y + z) = 18$$

Adding corresponding sides of the above equations, we have

81

$$(x + y + z)^2 = 36,$$

so that $x + y + z = \pm 6$,

When $x + y + z = 6$, we have $x = -2$, $y = 5$, $z = 3$

When $x + y + z = -6$, we have $x = 2$, $y = -5$, $z = -3$

Hence the solutions are $x = -2$, $y = 5$, $z = 3$; or $x = 2$, $y = -5$, $z = -3$.

16.   Subtracting both sides of (2) from corresponding sides of (1), we have

$$y^2 - x^2 + z(y - x) = -20,$$

or $(y - x)(x + y + z) = -20$ ... (4)

Again, subtracting both sides of (3) from corresponding sides of (2), we have

$$z^2 - y^2 + x(z - y) = -10,$$

or $(z - y)(x + y + z) = -10,$ ... (5)

From (4) and (5), we have by division

$$\frac{y - x}{z - y} = 2,$$

or $x = 3y - 2z$,   ... (6)

Substituting the value of x from (6) in (2), we have

or $3y^2 - 3yz + z^2 = 13$,   ... (7)

Also, from (1), we have

$$y^2 + yz + z^2 = 19,$$

From (7) and (1), we have

$$19(3y^2 - 3yz + z^2) = 13(y^2 + yz + z^2),$$

or $44y^2 - 70yz + 6y^2 = 0$

i.e., $22y^2 - 35yz + 3z^2 = 0$

i.e., $22y^2 - 35yz + 3z^2 = 0$,

which gives $z = \dfrac{2}{3}y$ *or* $z = 11y$

When $z = \dfrac{2}{3}y$, (1) gives $y^2 = 9$, i.e., $y = \pm 3$, so that $z = \pm 2$ and (6) gives $x = \pm 5$.

When $z = 11y$, (1) gives $y^2 = \dfrac{1}{7}$, so that $y = \pm \dfrac{1}{\sqrt{7}}$.

Consequently $z = \pm \dfrac{11}{\sqrt{7}}$, and $x = \mp \dfrac{19}{\sqrt{7}}$

Thus the solutions are

$$x = \pm 5,\ y = \pm 3,\ z = \pm 2\ ;$$

or
$$x = \mp \dfrac{19}{\sqrt{7}},\ y = \pm \dfrac{1}{\sqrt{7}},\ z = \pm \dfrac{11}{\sqrt{7}}$$

17. From the second and third equations, we get

$$(x + y + z)^2 = a^2 = x^2 + y^2 + z^2$$

so that $xy + yz + zx = 0$.

From the first equation, we have

$$x^3 + y^3 + z^3 - 3xyz = a^3 - 3xyz,$$

so that $\quad (x + y + z)(x^2 + y^2 + z^2 - yz - zx - xy) = a^3 - 3xyz,$

i.e., $\quad a.a^2 = a^3 - 3xyz$

so that $\quad xyz = 0$.

18. Since $(a + b + c)^2 = a^2 + b^2 + c^2 + 2(bc + ca + ab)$, therefore

$$bc + ca + ab = [1^2 - 9]/2 = -4 \qquad \dots (1)$$

Also, $a^3 + b^3 + c^3 - 3abc = (a + b + c)(a^2 + b^2 + c^2 - bc - ca - ab)$

$$= 1.(9 - (-4))$$

$$= = 13, \qquad \dots (2)$$

so that $\quad abc = (1 - 13)/3 = -4$, $\qquad \dots (3)$

Now, $\quad \dfrac{1}{a} + \dfrac{1}{b} + \dfrac{1}{c} = (bc + ca + ab)/abc$

$$= (-4)/(-4) = 1$$

19.   The given equations can be re-written as

$$3x(x+y) = 6x + 2y$$

$$y(x+y) = 9x + y$$

Dividing the corresponding sides of the above equations, we have

either $x+y=0$, or $\dfrac{3x}{y} = \dfrac{6x+2y}{9x+y}$

Two different cases arise.

Case I. $x+y=0$. Then we get $6x+2y=0, 9x+y=0$ which give $x=0, y=0$ as a solution.

Case II. $\dfrac{3x}{y} = \dfrac{6x+2y}{9x+y}$

so that $3x(9x+y) = y(6x+2y)$

or $27x^2 - 3xy - 2y^2 = 0$

or $(3x-y)(9x+2y) = 0$

i.e., $y = 3x$ or $y = -\dfrac{9}{2}x$.

When $y=3x, 12x^2 = 12x,$ so that $x=0$ or 1.

If $x=0$, then $y=0$; if $x=1$, then $y=3$. Therefore $x=0, y=0$, and $x=1, y=3$ are solutions.

When, $y = -\dfrac{9}{2}x,$

$3x(x+y) = 6x+2y$ yields

$$3x\left(x - \dfrac{9}{2}x\right) = 6x + 2\left(-\dfrac{9}{2}x\right),$$

i.e., $3x\left(-\dfrac{7x}{2}\right) = -3x$,

i.e., $7x^2 = 2x$

84

i.e., $\qquad x = 0 \ or \ \dfrac{2}{7}$

If $x = 0, \ y = 0$; if $x = \dfrac{2}{7}, \ y = -\dfrac{9}{2}\left(\dfrac{2}{7}\right) = \dfrac{-9}{7}$

Collecting all the solutions together we find that $x = 0,$

$y = 0$; $x = 1, \ y = 3$ and $x = \dfrac{2}{7}, \ y = -\dfrac{9}{7}$ are all the solutions.

Aliter: From $3x(x + y) = 6x + 2y,$ we have

$\qquad y = (6x - 3x^2)/(3x - 2)$

we have $\quad \left[(6x - 3x^2)/(3x - 2)\right]\left[4x/(3x - 2)\right]$

$\qquad\qquad = 9x + (6x - 3x^2)/(3x - 2),$

i.e., $\qquad 4x(6x - 3x^2) = 9x(3x - 2)^2 + (3x - 2)(6x - 3x^2)$

i.e., $\qquad 3x\left[(x + 2)(2 - x) - 3(3x - 2)^2\right] = 0$

i.e., $\qquad 3x(-28x^2 + 36x - 8) = 0$

or $\qquad x(7x^2 - 9x + 2) = 0$

so that $\quad x = 0, 1 \ or \ \dfrac{2}{7}$

When $x = 0, \ y = 0$; when $x = 1, \ y = 3$; $x = \dfrac{2}{7}, \ y = -\dfrac{9}{7}$

Thus $\quad x = 0, \ y = 0$; $x = 1, \ y = 3$; $x = \dfrac{2}{7}, \ y = \dfrac{-9}{7}$ are the desired solutions.

20. $\quad \log_3(\log_2 x) + \log_{1/3}(\log_{1/2} y) = 1 \qquad\qquad \text{... (i)}$

Now, $\log_{1/2} y = (\log_2 y) \times (\log_{1/2} 2),$

$\qquad\qquad = -\log_2 y, \ \text{since } \log_{1/2} 2 = -1 \qquad\qquad \text{... (ii)}$

Also, $\log_{1/3}(\log_{1/2})y = \log_3(\log_{1/2} y) \times (\log_{1/3} 3),$

$\qquad\qquad = -\log_3(\log_{1/2} y), \ \text{since } \log_{1/3} 3 = -1,$

$$= -\log_3(-\log_2 y), \text{ by (ii)},\qquad\qquad \dots \text{(iii)}$$

From (i) and (iii), we have

$$\log_3(\log_2 x) - \log_3(-\log_2 y) = 1,$$

i.e., $\quad \log_3\big((-\log_2 x)/(\log_2 y)\big) = 1$

i.e., $\quad \big((-\log_2 x)/(\log_2 y)\big) = 3^1 = 3$

i.e., $\quad 3\log_2 y + \log_2 x = 0,$

or $\quad xy^3 = 1$

Also $\quad xy^2 = 4$

From the above we have $y = \dfrac{1}{4}, \ x = 64$

Therefore $x = 64, \ y = \dfrac{1}{4}$ is the desired solution.

21. $\log_2 x + \log_4 y + \log_4 z = 2$

$\Rightarrow \quad (\log_4 x)\log_2 4 + \log_4 y + \log_4 z = 2$

$\Rightarrow \quad x^2 yz = 16, \text{ since } \log_2 4 = 2$

Similarly the remaining equations reduced to

$$y^2 zx = 81 \ \ and \ \ z^2 xy = 256$$

Solving the equations $x^2 yz = 16, \ xy^2 z = 81, \ xyz^2 = 256$, we get

$x = \dfrac{2}{3}, \ y = \dfrac{27}{8}, \ z = \dfrac{32}{3}$, since x, y, z must be all positive.

22.
$$(x+y)(x+y+z) = 18,$$
$$(y+z)(x+y+z) = 30,$$
$$(z+x)(x+y+z) = 2A,$$

Adding corresponding sides of the above equations, we have

$$2(x+y+z)^2 = 48 + 2A$$

$\Rightarrow \quad x+y+z = \pm(24+A)^{\frac{1}{2}} = k \ (say).$

If $k = 0$, the equations do not have a solution. (In fact, the equations give inconsistent relations such as $0 = 18, 0 = 30$).

Let us therefore take $k \neq 0$. We then have

$$x + y = \frac{18}{k}, \ y + z = \frac{30}{k}, \ z + x = \frac{2A}{k}, \ \textit{so that}$$

$$x = k - \frac{30}{k} = (k^2 - 30)/k = (6 - A)/k$$

$$y = k - \frac{2A}{k} = (k^2 - 2A)/k = (24 - A)/k$$

$$z = k - \frac{18}{k} = (k^2 - 18)/k = (6 + A)/k$$

where $\quad k = \pm(24 + A)^{\frac{1}{2}}$.

23. $\quad (x - 4)(y - 4) = 16, \qquad \Rightarrow \qquad 4(x + y) = xy,$

$\quad (y - 6)(z - 6) = 36, \qquad \Rightarrow \qquad 6(y + z) = yz,$

$\quad (z - 8)(x - 8) = 64, \qquad \Rightarrow \qquad 8(z + x) = zx.$

If none of xy, yz, zx is zero, then

$$\frac{x + y}{xy} = \frac{1}{4}, \ \frac{y + z}{yz} = \frac{1}{6}, \ \frac{z + x}{zx} = \frac{1}{8}$$

i.e., $\quad \dfrac{1}{x} + \dfrac{1}{y} = \dfrac{1}{4}, \ \dfrac{1}{y} + \dfrac{1}{z} = \dfrac{1}{6}, \ \dfrac{1}{z} + \dfrac{1}{x} = \dfrac{1}{8} \qquad \dots (1)$

By adding the respective side of the above equations, we have

$$\frac{1}{x} + \frac{1}{y} + \frac{1}{z} = \frac{1}{2}\left(\frac{1}{4} + \frac{1}{6} + \frac{1}{8}\right) = \frac{13}{48} \qquad \dots (2)$$

From (1) and (2), we have

$$\frac{1}{x} = \frac{13}{48} - \frac{1}{6} = \frac{5}{48}, \ \frac{1}{y} = \frac{13}{48} - \frac{1}{8} = \frac{7}{48}, \ \frac{1}{z} = \frac{13}{48} - \frac{1}{4} = \frac{1}{48}$$

$$\therefore \qquad x = \frac{48}{5}, \ y = \frac{48}{7}, \ z = 48$$

87

If some one of xy, yx, zx is zero, then it is easy to see that x, y, z are all zero.

Thus the solutions are $x = y = z = 0$; $x = \dfrac{48}{5}$, $y = \dfrac{48}{7}$, $z = 48$.

24.   Initiate the method of the above example. The solutions are

$$x = y = z = 0;\ x = \frac{24}{5},\ y = \frac{24}{7},\ z = 24.$$

25.   The given system of equations can be written as

$$(x+1)(y+1) = -12$$

$$(y+1)(z+1) = -8$$

$$(z+1)(x+1) = 6.$$

Multiplying the corresponding sides of the equations and taking square roots, we have

$$(x+1)(y+1)(z+1) = \pm 24.$$

When we take the positive sign on the RHS, we get

$$x+1 = -3,\ y+1 = 4,\ z+1 = -2,$$

i.e., $(x, y, z) = (-4, 3, -3)$

When we take the negative sign on the RHS.

we get $x+1 = 3,\ y+1 = -4,\ z+1 = 2$.

i.e., $(x, y, z) = (2, -5, 1)$

Thus the solutions are

$$x = -4,\ y = 3,\ z = -3;$$

or $x = 2,\ y = -5,\ z = 1$

26.
$$(5x)^2 + (5y)^2 = \frac{12^2}{\left(1 + \dfrac{1}{x^2 + y^2}\right)^2} + \frac{4^2}{\left(1 - \dfrac{1}{x^2 + y^2}\right)^2}$$

Putting $x^2 + y^2 = \dfrac{1}{t}$, we have

$$\frac{25}{t} = \frac{144}{(1+t)^2} + \frac{16}{(1-t)^2}$$

so that $25(1-t^2)^2 = 144t(1-t)^2 + 16t(1+t)^2$,

or $\qquad 25t^4 - 160t^3 + 206t^2 - 160t + 25 = 0$

Dividing throughout by t^2, we get

$$25t^2 - 160t + 206 - \frac{160}{t} + \frac{25}{t^2} = 0,$$

i.e., $\qquad 25\left(t^2 + \frac{1}{t^2}\right) - 160\left(1 + \frac{1}{t}\right) + 206 = 0$

Putting $t + \frac{1}{t} = u$, we have

$$25(u^2 - 2) - 160u + 206 = 0$$

i.e., $\qquad 25u^2 - 160u + 156 = 0$

$\therefore \qquad u = \frac{160 \pm (160^2 - 4.25.156)}{50}$

$$= \frac{160 \pm 100}{50} = \frac{6}{5}, \frac{26}{5}$$

When $u = \frac{6}{5}$, $t + \frac{1}{t} = \frac{6}{5}$, which does not give any real values of t.

When $u = \frac{26}{5}$, $t + \frac{1}{t} = \frac{26}{5}$, which gives $t = \frac{1}{5}$ or 5.

$\therefore \qquad t = 5 \ or \ \frac{1}{5}$, so that $x^2 + y^2 = \frac{1}{5}$ or 5.

When $x^2 + y^2 = \frac{1}{5}$, we get

$$5x(1+5) = 12, \ 5y(1-5) = 4$$

so that $x = \frac{2}{5}, y = -\frac{1}{5}$

When $x^2 + y^2 = 5$, we get $5x\left(1 + \dfrac{1}{5}\right) = 12$, $5y\left(1 - \dfrac{1}{5}\right) = 4$

so that $x = 2$, $y = +1$.

Thus $x = 2/5$, $y = -1/5$ and $x = 2$, $y = +1$ are the desired solutions.

27. The given equations can be written as

$$x - z = 4 - y,$$

$$x^2 + z^2 = y^2 - 4,$$

$$xz = 6/y.$$

By using the identity $(x - z)^2 + 2xz = x^2 + z^2$, we eliminate x and z from the above equations, so as to get

$$(4 - y)^2 + \frac{12}{y} = y^2 - 4,$$

$\Rightarrow \quad y(4 - y)^2 + 12 = y(y^2 - 4) \quad \Rightarrow \quad 2y^2 - 5y - 3 = 0$

$\Rightarrow \quad y = -\dfrac{1}{2} \quad or \quad 3$.

When $y = -\dfrac{1}{2}$, $x^2 + z^2 = -\dfrac{15}{4}$ which is not possible for any real values of x and z.

When $y = 3$, $x - z = 1$, $xz = 2$, so that x and $-z$ are the roots of $t^2 - t - 2 = 0$ giving $t = 2$ or -1, i.e., either $x = 2$, $z = 1$ or $x = -1$, $z = -2$.

Hence, there are two solutions, namely $x = 2$, $y = 3$, $z = 1$ and $x = -1$, $z = -2$.

Hence there are two solutions, namely $x = 2$, $y = 3$, $z = 1$ and $x = -1$, $y = 2$, $z = -2$.

EXERCISE - 2   [Time: 3hrs]

1. Solve: $x + y = 3$, $\dfrac{x^2}{y} + \dfrac{y^2}{x} = \dfrac{9}{2}$

2. Solve: $3x^2 + 3xy + y^2 = 7$

$2x^2 - 3xy + 2y^2 = 14$

3. When x, y, z are real solve the equations $x + y = 2$ and $xy - z^2 = 1$.

4. If $x < 0$ and $y < 0$ solve the equation.

$$x + y + \frac{x}{y} = \frac{1}{2}$$

$$(x + y)\left(\frac{x}{y}\right) = -\frac{1}{2}$$

5. Solve: $x^4 + y^4 = 82$

$x + y = 4$

6. If $x = cy + bz$, $y = az + cx$ and $z = bx + ay$

where $x \neq 0$, $y \neq 0$, $z \neq 0$.

Prove that $a^2 + b^2 + c^2 + 2abc = 1$.

7. Solve $x^2 - yz = a^2$, $y^2 - zx = b^2$, $z^2 - xy = c^2$.

8. Solve $x + y + z = 9$

$x^2 + y^2 + z^2 = 29$

$x^3 + y^3 + z^3 = 99$

9. Solve $x^2 + y(x + 1)17$

$y^2 + x(y + 1) = 13$

EXERCISE - 2
Answers

1. $(1, 2), (2, 1)$
2. $(3, 4), (-3, -4)$
3. $x = 1, y = 1$ & $z = 0$

4. $-\frac{1}{4}, -\frac{1}{4}$

5. $x = 2 - 5i$, $y = 2 + 5i$

8. $2, 3, 4$

9. $x = -\frac{23}{7}, y = -\frac{19}{7}$

BLANK PAGE LEFT INTENTIONALLY

5. POLYNOMIALS

1. **Real Polynomial**

   Let a_0, a_1, a_2,........,a_n be real numbers and x is a real variable. Then,

   $$f(x) = a_0 + a_1 x + a_2 x^2 + ... + a_n x^n$$

   is called a real polynomial of real variable x with real coefficients.

2. **Degree of a Polynomial**

   The degree of a real polynomial is the highest power of x in the polynomial.

   $$f(x) = a_n x^n + a_{n-1} x^{n-1} + a_{n-2} x^{n-2} + ... + a_0 x^0$$

   is a polynomial of degree n as highest power of x where n is a positive integer and $a_n \neq 0$.

   Linear Polynomial: Polynomial of degree one is known as linear polynomial.

   Note: $f(x) = ax + b$ is a general polynomial of degree one known as linear polynomial $a \neq 0$.

   Quadratic Polynomial: Polynomial of degree two is known as quadratic polynomial.

   Note: $f(x) = ax^2 + bx + c$, $a \neq 0$ is a general of degree two polynomial known as quadratic polynomial.

3. **Polynomial Equation**

   If $y = f(x)$ is a real polynomial of degree n, then $f(x) = 0$ is a corresponding real polynomial equation of degree n.

4. **Roots of an Equation**

   Roots of an equation in x are those values of x which satisfy the equation.

   OR

   If $f(\alpha) = 0$, then $x = \alpha$ is the root of the equation $f(x) = 0$.

   Note: * Real roots of an equation $f(x) = 0$ are the x co-ordinates of the points where graph of $y = f(x)$ intersects X-axis

   * An equation of degree n has n roots, (not necessarily all real).

5. **Remainder Theorem:**

   If a polynomial $f(x)$ is divided by $x - \alpha$, the remainder obtained is

Problems in Algebra

$f(\alpha)$.

Factor Theorem:

A polynomial $f(x)$ is divisible by $x - \alpha$, if $f(\alpha) = 0$,

* If $\dfrac{a_1}{a_2} = \dfrac{b_1}{b_2} = \dfrac{c_1}{c_2}$, then each of these ratios is also equal to:

(i) $\dfrac{ka_1 + lb_1 + mc_1}{ka_2 + lb_2 + mc_2}$ (where $k, l, m \in R$)

(ii) $\left(\dfrac{ka_1^x + lb_1^x + mc_1^x}{ka_2^x + lb_2^x + mc_2^x}\right)^{1/x}$

For example: $\dfrac{\alpha}{\beta} = \dfrac{3}{2}$, then: $\dfrac{\alpha}{\beta} = \dfrac{3}{2} = \dfrac{\alpha+3}{\beta+2} = \sqrt{\dfrac{\alpha^2+9}{\beta^2+4}}$

6. Condition for Resolution into linear factors of a quadratic function

The quadratic function $ax^2 + 2hxy + by^2 + 2gx + 2fy + c$ is resolvable into linear rational factors iff

$$abc + 2fgh - af^2 - bg^2 - ch^2 = 0 \quad i.e. \quad \begin{vmatrix} a & h & g \\ h & b & f \\ g & f & c \end{vmatrix} = 0$$

7. Transformation of Equations

1. Transformation of an equation into another equation whose roots are the reciprocals of the roots of the given equation.

Let $\quad f(x) = a_0 x^n + a_1 x^{n-1} + a_2 x^{n-2} + \ldots\ldots a_{n-1} x + a_n = 0 \quad \ldots$ (i)

be the given equation. Let x and y be respectively the roots of the given equation and that of the transformed equation. Then,

$$y = \frac{1}{x} \quad \Rightarrow \quad x = \frac{1}{y}$$

Putting $x = \dfrac{1}{y}$ in (i), we get:

$$\frac{a_0}{y^n} + \frac{a_1}{y^{n-1}} + \frac{a_2}{y^{n-2}} + \dots + \frac{a_{n-1}}{y} + a_n = 0 \Rightarrow a_n y^n + a_{n-1} y^{n-1} + \dots + a_1 y + a_0 = 0$$

This is the required equation.

Note: Thus, an equation whose roots are reciprocal of the roots of a given equation x by $1/y$ in the given equation.

2. Transformation of an equation to another equation whose roots are negative of the roots of a given equation

Let the given equation be

$$f(x) = a_0 x^n + a_1 x^{n-1} + a_2 x^{n-2} + \dots + a_{n-1} x + a_n = 0$$

Note: Let x be a root of the given equation and y be a root of the transformed equation. Then, $y = -x \ or \ x = -y$. Thus, the transformed equation is obtained by putting $x = -y$ in $f(x) = 0$ and is therefore $f(-y) = 0$

or $\qquad a_0 y^n - a_1 y^{n-1} + a_2 y^{n-2} + \dots + (-1)^n a_n = 0$

3. Transformation of an equation to another equation whose roots are square of the roots of a given equation

Let x be a root of the given equation and y be that of the transformed equation. Then,

$$y = x^2 \Rightarrow x = \sqrt{y} .$$

Note: Thus, an equation whose roots are squares of the roots of a given equation is obtained by replacing x by $\sqrt{y}$ in the given equation.

4. Transformation of an equation to another equation whose roots are cubes of the roots of a given equation

Let x be a root of the given equation and y be that of the transformed equation. Then,

$$y = x^3 \Rightarrow x = y^{1/3}$$

Note: Thus, an equation whose roots are cubes of the roots of a given equation is obtained by replacing x by $y^{1/3}$ in the given equation.

8.   Relations between roots and coefficients

Problems in Algebra

If $\alpha_1, \alpha_2, \alpha_3, \ldots \alpha_n$ are roots of the equation

$$f(x) = a_0 x^n + a_1 x^{n-1} + a_2 x^{n-2} + \ldots\ldots + a_{n-1}x + a_n = 0$$

$$f(x) = a_0(x - \alpha_1)(x - \alpha_2)(x - \alpha_3)\ldots(x - \alpha_n)$$

$$\therefore \ a_0 x^n + a_1 x^{n-1} + a_2 x^{n-2} + \ldots.. + a_{n-1}x + a_n = a_0(x - \alpha_1)(x - \alpha_2)\ldots..(x - \alpha_n)$$

Comparing the coefficients of x^{n-1} on both sides, we get:

$$S_1 = \alpha_1 + \alpha_2 + \ldots.. + \alpha_n = \Sigma \alpha_i = \frac{-a_1}{a_0}$$

or, $$S_1 = -\frac{coeff. \ of \ x^{n-1}}{coeff. \ of \ x^n}$$

Comparing the coefficients of x^{n-2} on both sides, we get:

$$S_2 = \alpha_1\alpha_2 + \alpha_1\alpha_3 + \ldots.. = \sum_{i \neq j}\alpha_i\alpha_j = (-1)^2\frac{a_2}{a_0}$$

or, $$S_2 = \frac{(-1)^2 \ coeff. \ of \ x^{n-2}}{coeff. \ of \ x^n}$$

Comparing the coefficients of x^{n-3} on both sides, we get:

$$S_3 = \alpha_1\alpha_2\alpha_3 + \alpha_2\alpha_3\alpha_4 + \ldots.. = \sum_{i \neq j \neq k}\alpha_i \alpha_j \alpha_k = (-1)^3\frac{a_3}{a_0}$$

or, $$S_3 = \frac{(-1)^3 \ coeff. \ of \ x^{n-3}}{coeff. \ of \ x^n}$$

...   .

...   .

$$S_n = \alpha_1\alpha_2\alpha_3\ldots..\alpha_n = (-1)^n\frac{a_n}{a_0} = (-1)^n\frac{constant \ term}{coeff. \ of \ x^n}$$

Here, S_k denotes the sum of the products of the roots taken k at a time.

Particular Cases:

Quadratic Equation: If α, β are roots of the quadratic equation

$ax^2 + bx + c = 0,$ then

$$\alpha + \beta = -\frac{b}{a} \qquad and \qquad \alpha\beta = \frac{c}{a}$$

Cubic Equation: If α, β, γ are roots of a cubic equation

$$ax^3 + bx^2 + cx + d = 0, \text{ then } \alpha + \beta + \gamma = -\frac{b}{a},$$

$$\alpha\beta + \beta\gamma + \gamma\alpha = (-1)^2 \frac{c}{a} = \frac{c}{a} \quad and \quad \alpha\beta\gamma = (-1)^3 \frac{d}{a} = -\frac{d}{a}.$$

Biquadratic Equation: If $\alpha, \beta, \gamma, \delta$ are roots of the biquadratic equation $ax^4 + bx^3 + cx^2 + dx + e = 0$, then

$$S_1 = \alpha + \beta + \gamma + \delta = -\frac{b}{a}$$

$$S_2 = \alpha\beta + \beta\gamma + \alpha\delta + \alpha\gamma + \beta\delta + \gamma\delta = (-1)^2 \frac{c}{a} = \frac{c}{a}$$

or, $$S_2 = (\alpha + \beta)(\gamma + \delta) + \alpha\beta + \gamma\delta = \frac{c}{a}$$

$$S_3 = \alpha\beta\gamma + \beta\gamma\delta + \gamma\delta\alpha + \alpha\beta\delta = (-1)^3 \frac{d}{a} = -\frac{d}{a}$$

or, $$S_3 = \alpha\beta(\gamma + \delta) + \gamma\delta(\alpha + \beta) = -\frac{d}{a}$$

and, $$S_4 = \alpha\beta\gamma\delta = (-1)^4 \frac{e}{a} = \frac{e}{a}$$

9. Maximum number of positive (negative) roots of a polynomial:

Rule 1: The maximum number of positive real roots of a polynomial equation

$$f(x) = a_0 x^n + a_1 x^{n-1} + a_2 x^{n-2} + \dots + a_{n-1}x + a_n = 0$$

is the number of changes of the signs of coefficients from positive to negative or negative to positive.

For example, in the equation $x^3 + 3x^2 + 7x - 11 = 0,$ the signs of coefficients are

97

$$+ + + -$$

As there is just one change of sign, the number of positive roots of $x^3 + 3x^2 + 7x - 11 = 0$ is at most 1.

Rule 2: The maximum number of negative roots of the polynomial equation $f(x) = 0$ is the number of changes of the signs of coefficients from positive to negative or negative to positive of the equation $f(-x) = 0$.

EXERCISE - 1

1.   Find all rational numbers a, b, c such that the equation $x^3 + ax^2 + bx + c = 0$ has roots a, b, c.

2.   Find integers 'a' and 'b' such that $(x^2 - x - 1)$ divides $ax^{17} + bx^{16} + 1$.

3.   Show that there do not exist polynomials $p(x)$ and $q(x)$ each having integer coefficients and degree greater than or equal to 1 such that $p(x)q(x) = x^5 + 2x + 1$.

4.   The polynomial $1 - x + x^2 - x^3 + \ldots\ldots + x^{16} - x^{17}$ may be written in the form $a_0 + a_1 y + a_2 y^2 + \ldots\ldots + a_{16} y^{16} + a_{17} y^{17}$

where $y = x + 1$ and $a_i s$ are constants. Find a_2.

5.   Find polynomials $f(x), g(x)$ and $h(x)$. If they exist, such that for all x,

$$|f(x)| - |g(x)| + h(x) = \begin{cases} -1 & \text{if } x < -1 \\ 3x + 2 & \text{if } -1 \le x \le 0 \\ -2x + 2 & \text{if } x > 0 \end{cases}$$

6.   Find the real zeros of the polynomial

$$P_a(x) = (x^2 + 1)(x - 1)^2 - ax^2$$

where a is a given real number.

7.   Solve the equation $x^3 - 6x^2 + 11x - 6 = 0$, the roots being in A.P.

8.   Solve the equation $27x^3 + 42x^2 - 28x - 8 = 0$, the roots being in G.P.

9.   Solve the equation $3x^3 + 11x^2 + 12x + 4 = 0$, being given that the roots are in H.P.

10. Solve the equation

$$3x^4 - 40x^3 + 130x^2 - 120x + 27 = 0,$$

given that the product of two of its roots is equal to the product of the other two.

11. Solve the equation

$$x^4 + 15x^3 + 70x^2 + 120x + 64 = 0$$

the roots being in G.P.

12. If α, β, γ, δ be the roots of the equation

$$x^4 + px^3 + qx^2 + rx + s = 0,$$

show that $(1+\alpha^2)(1+\beta^2)(1+\gamma^2)(1+\delta^2) = (1-q+s)^2 + (p-r)^2$.

13. If $1, \alpha_1, \alpha_2, \ldots, \alpha_{n-1}$ be the roots of the equation $x^n - 1 = 0$, show that $n = (1-\alpha_1)(1-\alpha_2)(1-\alpha_3)\ldots(1-\alpha_{n-1})$.

14. If
$$\alpha + \beta + \gamma = 1,$$
$$\alpha^2 + \beta^2 + \gamma^2 = 2,$$
$$\alpha^3 + \beta^3 + \gamma^3 = 3$$

find the value of $\alpha^4 + \beta^4 + \gamma^4$.

15. Solve:
$$(x+y)(x+z) = 15,$$
$$(y+z)(y+x) = 18,$$
$$(z+x)(z+y) = 30,$$

16. Solve: $x^2 + xy + xz = -12,$
$$y^2 + yz + yx = 30,$$
$$z^2 + zx + zy = 18,$$

17. Solve:
$$x^3 + y^3 + z^3 = a^3, \quad x^2 + y^2 + z^2 = a^2, \quad x+y+z = a$$

18. Find the remainder when $(x+1)^n$ is divided by $(x-1)^3$.

19. Solve the equation $x^4 - 4x^3 + x^2 - 2x + 1 = 0$.

20. $f(x)$ is a polynomial of degree atleast two with integer co-efficients.

Problems in Algebra

Show that which it is divided by $(x-a)(x-b)$, where $a \neq b$, the remainder is

$$x \left[\frac{f(a) - f(b)}{a-b} \right] + \frac{af(b) - bf(a)}{a-b}.$$

21. Factorise $x^6 + 5x^3 + 8$

22. Solve: $\left(\dfrac{x}{x+3} \right)^3 = \dfrac{x-3}{x+6}$

23. If -2 and 3 are two roots of the equation $x^4 + ax^2 + bx + a = 0$, find the value of b.

24. The product of two of the four roots of

$x^4 - 20x^3 + kx^2 + 590x - 1992 = 0$ is 24. Find k.

EXERCISE - 1
Solution

1. a, b, c are the roots of $x^3 + ax^2 + bx + c = 0$. We have,

$$a + b + c = -a \qquad \qquad \text{... (1)}$$
$$ab + bc + ca = b \qquad \qquad \text{... (2)}$$
$$abc = -c \qquad \qquad \text{... (3)}$$

Case (1) If $c = 0 (1) \Rightarrow b = -2a$ and $(2) \Rightarrow ab = b$ or $b(a-1) = 0$ or $a(a-1) = 0 \Rightarrow a = 0$ or $a = 1$. If $a = 0$, then $b = 0$ and if $a = 1$ then $b = -2$.

Thus we have $a = 0, b = 0, c = 0$. The equation is $x^3 = 0$. The roots are $0, 0, 0$.

If $a = -1, b = -2, c = 0$ then the equation becomes

$$x^3 + x^2 - 2x = 0 \quad (i.e.,) \quad x(x^2 + x - 2) = 0$$

(i.e.,) $x(x+2)(x-1) = 0$

The roots are $0, 1, -2$.

Case (2) If $c \neq 0 : (3) \Rightarrow ab = -1$.

100

Polynomials

$(2) \Rightarrow -1 + c(a+b) = b.\ (1) \Rightarrow c = -2a - b$.

Eliminating c and b we get

$2a^4 - 2a^2 - a + 1 = 0 \Rightarrow (a-1)(2a^3 + 2a^2 - 1) = 0$

If $a = 1$ then $b = -1$ and $c = -1$.

The equation becomes $x^3 + x^2 - x - 1 = 0$ whose roots are $1, -1, -1$.

Consider $2a^3 + 2a^2 - 1 = 0$. Let $\dfrac{p}{q}$ be a rational root $\Rightarrow$ p divides 1

and q divides 2. Thus the only possible roots are $\pm 1, \pm \dfrac{1}{2}$. But none

of these satisfies $2a^3 + 2a^2 - 1 = 0$.

$\therefore \qquad (a, b, c) = (0, 0, 0), (1, -2, 0), (1, -1, -1)$

2.   The divisor is $(x^2 - x - 1)$

Consider the equation $x^2 - x - 1 = 0$

$$x = \frac{1 \pm \sqrt{1+4}}{2} = \frac{1 \pm \sqrt{5}}{2}$$

Thus $p = \dfrac{1+\sqrt{5}}{2}$ and $q = \dfrac{1-\sqrt{5}}{2}$ are the roots.

These must be roots of $ax^{17} + bx^{16} + 1$ also

$\qquad$ (i.e.,) $\quad ap^{17} + bp^{16} + 1 = 0$

$\qquad\qquad\qquad aq^{17} + bq^{16} + 1 = 0$

We have $ap^{17} + bp^{16} = -1$ $\qquad\qquad$ ... (1)

and $\quad aq^{17} + bq^{16} = -1$ $\qquad\qquad$ ... (2)

$(1) \times q^{16}$ and $(2) \times p^{16}$ and using $pq = -1$ we get

$\qquad\qquad ap + b = -q^{16}$

$\qquad\qquad aq + b = -p^{16}$

$\Rightarrow a = \dfrac{p^{16} - q^{16}}{p - q} = (p^8 + q^8)(p^4 + q^4)(p^2 + q^2)(p+q)$ $\qquad$ ... (3)

101

Problems in Algebra

Since $p+q=1$ and $pq=-1$.

$$p^2+q^2=(p+q)^2-2pq=3$$

$$p^4+q^4=(p^2+q^2)^2-2p^2q^2=7$$

$$p^8+q^8=(p^4+q^4)^2-2p^4q^4=47$$

$(3)\Rightarrow a=(47)(7)(3)(1)=987$. Also,

$$a(p+q)+2b=-c(p^{16}+q^{16})=-\left[(p^8+q^8)^2-2p^8q^8\right]$$

$$=-2207$$

$\therefore$ $\qquad\qquad 2b=-2207-987=-3194$

$$b=-1597$$

3. Since the only possible integer roots of $x^5+2x+1=0$ are ±1 and these are not the zeros of this polynomial, it has no linear factors, so we can express $x^5+2x+1=p(x)q(x)$ where p is of degree 2 and q of degree 3. Since the coefficient of x^5 is 1 we can assume without loss of generality

$$p(x)=x^2+ax+b, \qquad q(x)=x^3+cx^2+dx+e$$

since $be=1$ either $b=e=1$ or $b=e=-1$

Case (i) Let $b=e=1$

We have $x^5+2x+1=(x^2+ax+b)(x^3+cx^2+dx+e)$

Comparing x terms we get

$$ae+bd=2 \qquad (i.e.,)\quad a+d=2 \qquad\qquad\dots(1)$$

Comparing x^4 and x^3 terms we get $a+c=0$ and

$$d+ac+b=0$$

$\Rightarrow c=-a \qquad and \qquad b+d=a^2\Rightarrow a^2=d+1$.

4. Let $f(x)$ denote the given expression. Then

$$xf(x)=x-x^2+x^3-\dots-x^{18}$$

and $\qquad (1+x)f(x)=1-x^{18}$

102

Hence $\quad f(x) = f(y-1) = \dfrac{1-(y-1)^{18}}{1+(y-1)} = \dfrac{1-(y-1)^{18}}{y}$

Therefore a_2 is equal to the coefficient of y^3 in the expansion of $1-(y-1)^{18}$

i.e., $\quad a_2 = {}^{18}C_3 = 816$.

5. Since $x = -1$ and $x = 0$ are the two critical values of the absolute functions, one can suppose that

$$F(x) = a|x+1| + b|x| + cx + d$$

$$= \begin{cases} (c-a-b)x+d-a & \text{if } x < -1 \\ (a+c-b)x+a+d & \text{if } -1 \leq x \leq 0 \\ (a+b+c)x+a+d & \text{if } x > 0 \end{cases}$$

which implies that $a = 3/2,\ b = -5/2,\ c = -1\ and\ d = 1/2$.

Hence $f(x) = (3x+3)/2$, $g(x) = 5x/2$, and $h(x) = -x + \dfrac{1}{2}$

6. We have $(x^2 + 1)(x^2 - 2x + 1) - ax^2 = 0$

Dividing by x^2 yields $\quad \left(x + \dfrac{1}{x}\right)\left(x - 2 + \dfrac{1}{x}\right) - a = 0$

By setting $y = x + 1/x$, the last equation becomes

$$y^2 - 2y - a = 0$$

It follows that $\quad x + \dfrac{1}{x} = 1 \pm \sqrt{1+a}$

which in turns implies that, if $a \geq 0$, then the polynomial $P_a(x)$ has the real zeros

$$x_{1,2} = \frac{1 + \sqrt{1+a} \pm \sqrt{a + 2\sqrt{1+a} - 2}}{2}$$

In addition, if $a \geq 8$, then $P_a(x)$ also has the real zeros

103

Problems in Algebra

$$x_{3,4} = \frac{1-\sqrt{1+a} \pm \sqrt{a-2\sqrt{1+a}-2}}{2}$$

7. Since the roots are in A.P., let them be denoted by $a-d,\ a,\ a+d$.

    Then, $\qquad \sigma_1 = 3a = 6$

    $$\sigma_2 = 3a^2 - a^2 = 11$$

    $$\sigma_3 = a(a^2 - a^2) = 6$$

    Solving for a and b, we have $a = 2,\ d = \pm 1$

    Hence the roots are 1, 2, 3.

8. Since the roots are in G.P., let them be denoted by $a/r,\ a,\ ar$.

    Then, $\quad \sigma_1 \equiv a/r + a + ar = -14/9$ $\qquad$ ... (i)

    $$\sigma_2 \equiv a(a/r + a + ar) = -28/27 \qquad \text{.... (ii)}$$

    $$\sigma_3 = a^3 = 8/27 \qquad \text{... (iii)}$$

    Dividing (ii) by (i), we have

    $$a = 2/3 \qquad \text{... (iv)}$$

    Substituting the value of a in (i), we have

    $$1/r + 1 + r = -7/3,$$

    or $\qquad 3r^2 + 10r + 3 = 0,$

    or $\qquad (3r+1)(r+3) = 0,$

    Thus, $\qquad r = -\dfrac{1}{3}\ or\ -3$ $\qquad$ ... (v)

    From (iv) and (v), we find that the roots are –2/9, 2/3, –2.

9. Let the roots of the given equation be $\alpha,\ \beta,\ \gamma$.

    Then, $\quad \sigma_1 \equiv \alpha + \beta + \gamma = -\dfrac{11}{3},$ $\qquad$ ... (i)

    $$\sigma_2 \equiv \beta(\alpha + \gamma) + \alpha\gamma = 4, \qquad \text{... (ii)}$$

    $$\sigma_3 = \alpha\beta\gamma = -4/3 \qquad \text{... (iii)}$$

    Also, since $\alpha,\ \beta,\ \gamma$ are in H.P.,

104

$$\therefore \qquad \beta = \frac{2\alpha\gamma}{\alpha + \gamma},$$

or $\qquad \beta(\alpha + \gamma) = 2\alpha\gamma$

or $\qquad \alpha\beta + \beta\gamma + \gamma\alpha = 3\alpha\gamma \qquad \qquad ...\text{(iv)}$

From (ii) and (iv), we have

$$3\alpha\gamma = 4, \qquad \qquad ...\text{(v)}$$

or $\qquad \alpha\gamma = \dfrac{4}{3}$

Dividing (iii) by (v), we have

$$\beta = -1 \qquad \qquad ...\text{(vi)}$$

Thus -1 is a root of the given equation, and the equation may be written as

$$(x + 1)(3x^2 + 8x + 4) = 0 \qquad \qquad ...\text{(vii)}$$

Solving $3x^2 + 8x + 4 = 0$, we find that the other two roots are

$$\frac{-8 \pm \sqrt{(64 - 48)}}{6} = -2, -\frac{2}{3}$$

Hence the roots are $-2, -1, -\dfrac{2}{3}$.

Remark. To write the given equation in the form (vii), we may first divide $3x^3 + 11x^2 + 12x + 4$ by $x + 1$.

10. Let the roots of the given equation be $\alpha, \beta, \gamma, \delta$.

Then, $\qquad \sigma_1 = (\alpha + \beta) + (\gamma + \delta) = \dfrac{40}{3} \qquad \qquad ...\text{(i)}$

$$\sigma_2 = (\alpha + \beta)(\gamma + \delta) + \alpha\beta + \gamma\delta = \frac{130}{3}, \qquad \qquad ...\text{(ii)}$$

$$\sigma_3 = \alpha\beta(\gamma + \delta) + \gamma\delta(\alpha + \beta) = 40 \qquad \qquad ...\text{(iii)}$$

$$\sigma_4 = \alpha\beta \cdot \gamma\delta = 9 \qquad \qquad ...\text{(iv)}$$

Since the product of two of the roots is equal to the product of the other two, therefore,

$$\alpha\beta = \gamma\delta \qquad \text{... (v)}$$

From (iii) and (v), we have

$$\alpha\beta(\alpha + \beta + \gamma + \delta) = 40 \qquad \text{... (vi)}$$

From (i), (v) and (vi), we have

$$\alpha\beta = \gamma\delta = 3 \qquad \text{... (vii)}$$

From (ii) and (vii), we have

$$(\alpha + \beta)(\gamma + \delta) = \frac{112}{3} \qquad \text{... (viii)}$$

From (i) and (viii), we find that $\alpha + \beta$ and $\gamma + \delta$ are the roots of the equation

$$t^2 - \frac{40}{3}t + \frac{112}{3} = 0$$

Solve the above equation, we have

$$t = 4,\ 28/3\ .$$

Therefore, $\alpha + \beta,\ \gamma + \delta = 4,\ 28/3$ \qquad ... (ix)

From (vii) and (ix), we find that two of the numbers $\alpha,\ \beta,\ \gamma,\ \delta$ are the roots of the equation

$$y^2 - 4y + 3 = 0 \qquad \text{... (x)}$$

and the remaining two are the roots of the equation

$$y^2 - (28/3)y + 3 = 0 \qquad \text{... (xi)}$$

Solving (x) and (xi), we find that the roots of the given equation are

$$1,\ 3,\ \frac{1}{3},\ 9\ .$$

11.   Let us first observe that if four numbers be in G.P., then the product of the first and the fourth is equal to the product of the second and the third. Therefore, if the roots are in G.P., the product of two of the roots must be equal to the product of the other two.

Let the roots be $\alpha,\ \beta,\ \gamma,\ \delta$. Then,

$$\alpha\beta = \gamma\delta \qquad \text{... (i)}$$

$$\sigma_1 = (\alpha + \beta) + (\gamma + \delta) = -15 \qquad \text{... (ii)}$$

$$\sigma_2 = (\alpha + \beta)(\gamma + \delta) + \alpha\beta + \gamma\delta = 70 \qquad \text{... (iii)}$$

$$\sigma_3 = \alpha\beta(\gamma + \delta) + \gamma\delta(\alpha + \beta) = -120 \qquad \text{... (iv)}$$

$$\sigma_4 = \alpha\beta.\gamma\delta = 64 \qquad \text{... (v)}$$

From (i), (ii) and (iv), we have

$$\alpha\beta = \gamma\delta = 8 \qquad \text{... (vi)}$$

From (iii) and (vi), we have

$$(\alpha + \beta)(\gamma + \delta) = 54 \qquad \text{... (vii)}$$

From (ii) and (vii), we find that $\alpha + \beta$ and $\gamma + \delta$ are the roots of the equation

$$t^2 + 15t + 54 = 0$$

$\therefore \qquad \alpha + \beta, \gamma + \delta = -6, -9 \qquad \text{... (viii)}$

From (vi) and (viii), we find that two of the numbers $\alpha, \beta, \gamma, \delta$ are the roots of the equation

$$y^2 + 9y + 8 = 0 \qquad \text{... (ix)}$$

and the other two are the roots of the equation

$$y^2 + 9y + 8 = 0 \qquad \text{... (x)}$$

Solving (ix) and (x), we find that the roots of the given equation are $-1, -8, -2, -4$.

Remark. We could have proceeded directly by taking the roots to be of the form $a/r^3, a/r, ar, ar^3$. But then we would find that it would not prove to be anything simpler than the above. The reader would do well to try it and convince himself about this observation.

12. Since $\alpha, \beta, \gamma, \delta$ be the roots of the equation

$$x^4 + px^3 + qr^2 + rx + s = 0 ,$$

therefore, $x^4 + px^3 + qx^2 + rx + s \equiv (x - \alpha)(x - \beta)(x - \gamma)(x - \delta)$.

Substituting $x = i, -i$ successively, we have

$$(1 - q + s) - i(p - r) = (i - \alpha)(i - \beta)(i - \gamma)(i - \delta), \qquad \text{... (i)}$$

$$(1 - q + s) + i(p - r) = (-i - \alpha)(-i - \beta)(-i - \gamma)(-i - \delta) \qquad \text{... (ii)}$$

Multiplying corresponding sides of (i) and (ii), we have

$$(1-q+s)^2 + (p-r)^2 = (1+\alpha^2)(1+\beta^2)(1+\gamma^2)(1+\delta)^2.$$

13.   Since, $1, \alpha_1, \alpha_2, \alpha_3, ..., \alpha_{n-1}$ are the roots of the equation

$$x^n - 1 = 0,$$

therefore $x^n - 1 \pm (x-1)(x-\alpha_1)(x-\alpha_2).....(x-\alpha_{n-1})$.

Dividing both sides of the above identity by $(x-1)$, we have

$$x^{n-1} + x^{n-2} + ... + 1 \pm (x-\alpha_1)(x-\alpha_2).....(x-\alpha_{n-1}) \qquad ... \text{(i)}$$

Since (i) is the true for all values of x, we have by putting $x = 1$.

$$n = (1-\alpha_1)(1-\alpha_2).....(1-\alpha_{n-1})$$

14.   We shall first determine the equation whose roots are α, β, γ and then find the sum of the fourth powers of the roots of the same.

Let the equation whose roots are α, β, γ be

$$x^3 + px^2 + qx + r = 0 \qquad ... \text{(i)}$$

Then, $\alpha + \beta + \gamma = -p$, $\qquad ... \text{(ii)}$

$\alpha\beta + \beta\gamma + \gamma\alpha = q$, $\qquad ... \text{(iii)}$

$\alpha\beta\gamma = -r$ $\qquad ... \text{(iv)}$

Substituting the value of $\alpha + \beta + \gamma$ in (ii), we have

$$p = -1 \qquad ... \text{(v)}$$

Also, by substituting the values of $\alpha + \beta + \gamma$ and $\alpha^2 + \beta^2 + \gamma^2$ in the identify

$$(\alpha + \beta + \gamma)^2 = \Sigma\alpha^2 + 2\Sigma\alpha\beta,$$

we have $q = \Sigma\alpha\beta = -\dfrac{1}{2}$ $\qquad ... \text{(vi)}$

Since α, β, γ are the roots of (i), by substituting α, β, γ for x in (i) in succession and adding, we have

$$S_3 + pS_2 + qS_1 + 3r = 0 \qquad ... \text{(vii)}$$

Substituting $S_1 = 1$, $S_2 = 2$, $S_3 = 3$, $p = -1$ and $q = -\dfrac{1}{2}$ in (vii)

we have

$$x^3 - x^2 - \frac{1}{2}x - \frac{1}{6} = 0 \qquad \qquad \text{... (ix)}$$

Multiplying (ix) throughout by x, we have

$$x^4 - x^3 - \frac{1}{2}x^2 - \frac{1}{6}x = 0 \qquad \qquad \text{... (x)}$$

Since α, β, γ satisfy (x), therefore, by substituting α, β, γ for x in (x) and adding, we get

$$S_4 - S_3 - \left(\frac{1}{2}\right)S_2 - \left(\frac{1}{6}\right)S_1 = 0 .$$

or $\qquad S_4 = S_3 + \frac{1}{2}S_2 + \frac{1}{6}S_1, = 3 + \frac{1}{2}.2 + \frac{1}{6}.1 = \frac{25}{6}.$

15. Let $y + z = X, z + x = Y, x + y = Z$.

Then, $YZ = 15, ZX = 18, XY = 30$. $\qquad$ ... (i)

Multiplying the corresponding sides of these equations, we have

$$X^2 Y^2 Z^2 = 30 \times 18 \times 15 = (15 \times 6)^2$$

$\therefore \qquad XYZ = \pm 90$.

When $XYZ = 90$, we get using equations (1),

$$X = 6, Y = 5, Z = 3,$$

i.e., $\qquad y + z = 6, z + x = 5, x + y = 3,$

so that $\qquad x = 1, y = 2, z = 4$

When $\quad XYZ = -90$, we get using equations (i), $X = -6, Y = -5, Z = -3$.

i.e., $\qquad y\ y + z = -6, z + x = -5, x + y = -3$.

so that $\qquad x = -1, y = -2, z = -4$.

Hence the solutions are

$$x = 1, y = 2, z = 4 .$$

or $\qquad x = -1, y = -2, z = -4$.

16. The given equations can be re-written as

$$x(x+y+z) = -12,$$

$$y(x+y+z) = 30,$$

$$z(x+y+z) = 18,$$

Adding corresponding sides of the above equations, we have

$$(x+y+z)^2 = 36,$$

so that $x+y+z = \pm 6$.

When $x+y+z = 6$, we have $x = -2, y = 5, z = 3$.

When $x+y+z = -6$, we have $x = 2, y = -5, z = -3$

Hence the solutions are $x = -2, y = 5, z = 3; x = 2, y = -5, z = -3$.

17.   From the second and third equations, we get

$$(x+y+z)^2 = a^2 = x^2 + y^2 + z^2,$$

so that $xy + yz + zx = 0$.

From the first equation, we have

$$x^3 + y^3 + z^3 - 3xyz = a^3 - 3xyz,$$

so that $(x+y+z)(x^2 + y^2 + z^2 - yz - zx - xy) = a^3 - 3xyz,$

i.e., $a. a^2 = a^3 - 3xyz,$

so that $xyz = 0$,

$\therefore$ x, y, z are the roots of the equation

$$t^3 - t^2 a = 0$$

$\therefore$ $t = 0, 0, a$, which gives

$x = 0, y = 0, z = a; x = 0, y = a, z = 0$; and

$x = a, y = 0, z = 0$ as the solutions.

18.   Writing $x - 1 = y$ (so that $x = 1 + y$), we find that $(x+1)^n = (2+y)^n$,

$$= 2^n + ny2^{n-1} + \frac{n(n-1)}{2} y^2 . 2^{n-2} + \dots \dots$$

The remainder, when the RHS is divided by y^3.

$$= 2^n + ny.2^{n-1} + \frac{n(n-1)}{2} y^2 2^{n-2},$$

$$= 2^n + n(x-1)2^{n-1} + \frac{n(n-1)}{2}(x-1)^2 . 2^{n-2},$$

$$= n(n-1).2^{n-3} x^2 + x[-n(n-1).2^{n-2} + n.2^{n-1}]$$

$$+ n(n-1).2^{n-3} - n.2^{n-1} + 2^n$$

$$= n(n-1).2^{n-3} x^2 - 2^{n-2} x[n^2 - 3n]$$

$$+ 2^{n-3}[n^2 - 5n + 8]$$

19. $x^4 - 2x^3 + x^2 - 2x + 1 = 0$,

$\Rightarrow \quad x^4 + 1 - 2x(x^2 + 1) + x^2 = 0$

$\Rightarrow \quad x^2 + \dfrac{1}{x^2} - 2\left(x + \dfrac{1}{x}\right) + 1 = 0$, on dividing throughout by x^2

$\Rightarrow \quad \left(x + \dfrac{1}{x}\right)^2 - 2\left(x + \dfrac{1}{x}\right) - 1 = 0$,

$\Rightarrow \quad x + \dfrac{1}{x} = \dfrac{2 \pm \sqrt{8}}{2} = 1 \pm \sqrt{2}$

Now $x + \dfrac{1}{x} = 1 + \sqrt{2} \Rightarrow x^2 - \left(\sqrt{2} + 1\right)x + 1 = 0$

$\Rightarrow \quad x = \dfrac{\left(\sqrt{2}+1\right) \pm \sqrt{\left\{\left(\sqrt{2}+1\right)^2 - 4\right\}}}{2},$

$\Rightarrow \quad x = \dfrac{\left(\sqrt{2}+1\right) \pm i\sqrt{\left(2\sqrt{2}-1\right)}}{2}$

Also $x + \dfrac{1}{x} = 1 - \sqrt{2} \Rightarrow x^2 + \left(\sqrt{2}-1\right)x + 1 = 0$

$\Rightarrow \quad x = \dfrac{-\left(\sqrt{2}-1\right) \pm \sqrt{\left\{\left(\sqrt{2}-1\right)^2 - 4\right\}}}{2}$

111

Problems in Algebra

$$\Rightarrow \qquad x = \frac{\left(-\sqrt{2}+1\right)\pm i\sqrt{\left(2\sqrt{2}+1\right)}}{2}$$

Hence the roots are

$$\frac{\left(\sqrt{2}+1\right)\pm\sqrt{\left(2\sqrt{2}-1\right)}}{2}, \quad \frac{\left(-\sqrt{2}+1\right)\pm i\sqrt{\left(2\sqrt{2}+1\right)}}{2}$$

20. Since the divisor is a polynomial of degree 2, the remainder will be of the form $Ax+B$. By division algorithm, we have

$$f(x) = q(x)(x-a)(x-b) + Ax + b \qquad \qquad \dots (1)$$

where $q(x)$ is the quotient.

Substituting $x=a$, $x=b$ successively in (1), we have

$$f(a) = Aa + B, \ f(b) = Ab + B.$$

Solving for A and B, we have

$$A = [f(a)-f(b)]/(a-b), \ B = [af(b)-bf(a)]/(a-b).$$

21. The given expression can be written as

$$(x^2)^3 + (-x)^3 + (2)^3 - 3.x^2.(-x).2,$$

which is of the form $a^3+b^3+c^3-3abc$, with $a=x^2, b=-x, c=2$.
Therefore the desired factorisation is

$$(a+b+c)(a^2+b^2+c^2-bc-ca-ab),$$
$$= (x^2-x+2)(x^4+x^2+4+2x-2x^2+x^3)$$
$$= (x^2-x+2)(x^4+x^3-x^2+2x+4)$$

22. Putting $\dfrac{x}{x+3} = y$, we have $x = \dfrac{3y}{1-y}(y\neq 1)$, the given equation

becomes $y^4 - 2y^3 + 3y - 1 = 0$, which gives $(y+1)(y-1)^3 = 0$

Since $y\neq 1$, therefore $y=-1$, whence $x=-3/2$.

23. We shall use the factor theorem. If $x+2$ and $x-3$ are the factors of

$f(x) \equiv x^4 + ax^2 + bx + a$, then $f(-2)=0, f(3)=0$, so that

$16+5a-2b=0, \ 81+10a+3b=0$, which give $a=-6, b=-7$.

112

24. Let the roots be $\alpha, \beta, \gamma, \delta$. Then

$$(\alpha + \beta) + (\gamma + \delta) = 20$$

$$(\alpha + \beta)(\gamma + \delta) + \alpha\beta + \gamma\delta = k,$$

$$(\alpha + \beta)\gamma\delta + (\gamma + \delta)\alpha\beta = -590.$$

$$(\alpha\beta).(\gamma\delta) = -1992.$$

Also, $\alpha\beta = 24$.

Then, $\gamma\delta = -83$

Now $(\alpha + \beta) + (\gamma + \delta) = 20, -82(\alpha + \beta) + 24(\gamma + \delta) = -590$, so that $(\alpha + \beta) = 10 = (\gamma + \delta)$.

These values give $k = (\alpha + \beta)(\gamma + \delta) + \alpha\beta + \gamma\delta = 41$.

EXERCISE - 2   **[Time: 3hrs]**

1. Given that the equation

$$x^4 + px^3 + qx^2 + rx + s = 0$$

has four real, positive roots, prove that

(a) $pr - 16s \geq 0$,

(b) $q^2 - 36s \geq 0$,

with equality in each case holding if and only if the four roots are equal.

2. Prove that the polynomial

$$f(x) = x^4 + 26x^2 + 52x^2 + 78x + 1989$$

cannot be expressed as a product

$$f(x) = p(x)q(x),$$

where $p(x), q(x)$ are both polynomials with integral co-efficients and degree less than 4.

3. Let a, b, c, d be four real numbers, not all equal to zero. Prove that the zeros of the polynomial

$$f(x) = x^6 + ax^3 + bx^2 + cx + d$$

cannot be all real.

Problems in Algebra

4. Prove that $a_1, a_2, \ldots\ldots, a_n$ are all distinct, then the polynomial

$$(x-a_1)^2 \, (x-a_2)^2 \ldots\ldots (x-a_n)^2 + 1$$

can never be written as the product of two polynomials with integer co-efficients.

5. If $p(x)$ is a polynomial with integer co-efficients, and a, b, c three distinct integers, then show that it is impossible to have

$$p(a) = b, \ p(b) = c, \ p(c) = a \ .$$

6. Let $f(x)$ be a polynomial with integer co-efficients and suppose that for five distinct integers a_1, a_2, a_3, a_4, a_5 one has

$$f(a_1) = f(a_2) = f(a_3) = f(a_4) = f(a_5) = 2$$

Prove that $f(x) \neq 9$ for any integer n.

7. Let $p(x) = x^2 + ax + b$ be a quadratic polynomial in which a and b are integers. Given any integer n, show that there is an integer M such that $p(n)p(n+1) = p(M)$.

8. A polynomial $f(x)$ with rational co-efficients leaves remainder 15 when divided by $x - 3$ and remainder $2x + 1$ when divided by $(x-1)^2$. Find the remainder when $f(x)$ is divided by $(x-3)(x-1)^2$.

9. For every pair p, q of positive integers whose HCF is 1, show that $x^p - 1)(x^q - 1)$ divides $(x^{pq} - 1)(x - 1)$.

EXERCISE - 2
Answers

8. $2x^2 - 2x + 3$

6. FUNCTIONAL EQUATIONS

1. An equation in which unknowns are functions is called a functional equation. We are asked to find all functions satisfying some given relation (relations).

2. While solving a functional equation, we need to keep in mind the property of domain of the functions, their range and also the given conditions on the functions.

EXERCISE - 1

1. If $2f(x-1) - f\left(\dfrac{1-x}{x}\right) = x$, then find $f(x)$.

2. If $f(n+1) = \dfrac{2f(n)+1}{2}$, where $n = 1, 2, ...$ and $f(1) = 2$, then find $f(101)$.

3. Determine all functions $f = R \sim \{0, 1\} \to R$ (here R denote the set of real numbers) satisfying the functional relation

   $f(x) + f\left(\dfrac{1}{1-x}\right) = \dfrac{2(1-2x)}{x(1-x)}$, for $x \neq 0$ and $x \neq 1$.

4. If $f\left(2x + \dfrac{y}{8}, 2x - \dfrac{y}{8}\right) = xy$, then prove that for all m and n,

   $f(m, n) + f(n, m) = 0$.

5. If $f(x) + 2f(1-x) = x^2 + 2, \ \forall \ x \in R$ then find $f(x)$.

6. Let f be a function satisfying $2f(xy) = \{f(x)\}^y + \{f(y)\}^x$ and

   $f(1) = k \neq 1$, then find the value of $\displaystyle\sum_{r=1}^{n} f(r)$

7. If $f(x)$ be a polynomial function satisfying

   $f(x).f\left(\dfrac{1}{x}\right) = f(x) + f\left(\dfrac{1}{x}\right)$ and $f(4) = 65$

   Then find $f(6)$.

Problems in Algebra

8. The function f, defined by $f(x) = \dfrac{ax+b}{cx+d}$

   where a, b, c and d are non zero real numbers, has the properties

   $f(19) = 19$, $f(97) = 97$, and $f(f(x)) = x$, for all values of x, except

   $-\dfrac{d}{c}$.

   Find the range of f.

9. Let $f : N \times N \to N$ be a function such that $f(1, 1) = 1$.

   $f(m+1, n) = f(m, n) + m$ and

   $f(m, n+1) = f(m, n) - n$ for all $m, n \in N$

   find all pairs (p, q) such that $f(p, q) = 2012$.

10. Let f be a function defined on $[0, 1]$ such that

    $f(0) = f(1) = 1$ and $|f(a) - f(b)| < |a - b|$, for all $a \neq b$ in the interval $[0, 1]$.

    Prove that $|f(a) - f(b)| < \dfrac{1}{2}$

11. Let $f(x) = \dfrac{2}{4^x + 2} \ \forall \ x \in R$.

    Evaluate $f\left(\dfrac{1}{2011}\right) + f\left(\dfrac{2}{2011}\right) + \ldots\ldots + f\left(\dfrac{2010}{2011}\right)$

12. If $f(1) = 1$ and $f(n) = n + f(n-1)$ for all natural numbers $n \geq 2$, then

    show that $f(n) = \dfrac{n(n+1)}{2}$.

    find the value of $f(2010)$.

13. Let $f(3n) = n + f(3n-3)$ where n is a positive integer greater than 1, and $f(3n) = 1$ when $n = 1$. Find the value of $f(12)$.

14. Let $f(3n) = n^2 + f(3n-3)$ and

    (i) n is a positive integer greater than 1.

(ii) $f(3) = 1$.

15. Let $f(x) = 1 - f(x-1)$. Then express $f(x+1)$ in terms of $f(x-1)$.

16. Let $f = ax + b$, $g = cx + d$, $\forall x \in R$ a, b, c, d are real constants.

    (i) Find relations between coefficients so that $f(g) = x$.

    (ii) Show that, when $f(g) = x$, $f(g) \Rightarrow g(f)$.

17. If $f(x) = -x^n(x-1)^n$, then

    find $f(x^2) + f(x)f(x+1)$

18. Let $f(n) = n(n+1)$ where n is a natural number. Find the values of m and n such that $4(fn) = f(m)$ where m is a natural number.

EXERCISE - 1
Solutions

1. $2f(x-1) - f\left(\dfrac{1-x}{x}\right) = x$ ... (i)

    Replace x by $\dfrac{1}{x}$, we get

    $2f\left(\dfrac{1}{x} - 1\right) - f\left(\dfrac{1 - \frac{1}{x}}{1/x}\right) = \dfrac{1}{x}$

    $\Rightarrow \quad 2f\left(\dfrac{1-x}{x}\right) - f(x-1) = \dfrac{1}{x}$ ... (ii)

    Multiplying (i) by 2 and adding to equation (ii), we get

    $3f(x-1) = 2x + \dfrac{1}{x}$

    $\Rightarrow \quad f(x-1) = \dfrac{1}{3}\left(2x + \dfrac{1}{x}\right)$

    Replace x by $x+1$, to get

    $f(x) = \dfrac{1}{3}\left[2(1+x) + \dfrac{1}{1+x}\right]$

Problems in Algebra

2.
$$f(n+1) = \frac{2f(n)+1}{2}$$

$\Rightarrow \quad f(n+1) = f(n) + \frac{1}{2}$

$\Rightarrow \quad f(n+1) - f(n) = \frac{1}{2}$

Substitute, $\quad n = 1 \quad \Rightarrow \quad f(2) - f(1) = \frac{1}{2}$

$\quad n = 2 \quad \Rightarrow \quad f(3) - f(2) = \frac{1}{2}$

$\quad : \qquad \qquad \qquad :$

$\quad n = 100 \quad \Rightarrow \quad f(101) - f(100) = \frac{1}{2}$

Adding all the equations, we get

$$f(101) = f(1) + 100 \times \frac{1}{2}$$

$\Rightarrow \quad f(101) = 52$

3. Put $y = \frac{1}{1-x}$, then

$$f(x) + f(y) = 2\left(\frac{1}{x} - y\right) \qquad \qquad \text{... (i)}$$

$$\left[\because \frac{2(1-2x)}{x(1-x)} = 2\left\{ \frac{(1-x)}{x(1-x)} - \frac{x}{x(1-x)} \right\} = 2\left\{ \frac{1}{x} - y \right\} \right]$$

Similarly, put $z = \frac{1}{1-y}$, then $x = \frac{1}{1-z}$.

and we have

$$f(y) + f(z) = 2\left(\frac{1}{y} - z\right) \qquad \qquad \text{... (ii)}$$

and
$$f(z) + f(x) = 2\left(\frac{1}{z} - x\right) \qquad \text{... (iii)}$$

Adding (i) and (ii), we get

$$2f(x) + f(y) + f(z) = 2\left(\frac{1}{x} - x\right) - 2y + \frac{2}{z}$$

using equation (ii), we have

$$2f(x) = 2\left(\frac{1}{x} - x\right) - 2\left(y + \frac{1}{y}\right) + 2\left(z + \frac{1}{z}\right)$$

$$\because \qquad y + \frac{1}{y} = \frac{1}{1-x} + 1 - x$$

$$z + \frac{1}{z} = \frac{x-1}{x} + \frac{x}{x-1}$$

$$\Rightarrow \qquad f(x) = \frac{x+1}{x-1}$$

$$\therefore \qquad f(x) = \frac{x+1}{x-1} \text{ is the only function satisfying the given}$$

functional equation.

4. $f\left(2x + \dfrac{y}{8}, \ 2x - \dfrac{y}{8}\right) = xy$

Let $\qquad 2x + \dfrac{y}{8} = \alpha \qquad$ and $\qquad 2x - \dfrac{y}{8} = \beta \qquad$ ... (i)

$$\Rightarrow \qquad x = \frac{\alpha + \beta}{4} \qquad \text{and} \qquad y = 4(\alpha - \beta)$$

Given, $\quad f\left(2x + \dfrac{y}{8}, \ 2x - \dfrac{y}{8}\right) = xy$

$$\Rightarrow \qquad f(\alpha, \beta) = \alpha^2 - \beta^2$$

$$\Rightarrow \qquad f(m, n) + f(n, m) = m^2 - n^2 + n^2 - m^2$$

$$\Rightarrow \qquad f(m, n) + f(n, m) = 0 \ \forall \ m, n$$

Problems in Algebra

5. Given $f(x) + 2f(1-x) = x^2 + 2$ ... (i)

Replace x by $1-x$ in equation (i), we get

$$f(1-x) + 2f(x) = (1-x)^2 + 2 \quad \text{... (ii)}$$

Now, multiplying equation (i) by 1 and equation (ii) by 2, then subtracting each other, we get

$$-3f(x) = x^2 + 2 - 2(1-x)^2 - 4$$

$$\Rightarrow \quad 3f(x) = x^2 - 4x + 4$$

$$\Rightarrow \quad f(x) = \frac{(x-2)^2}{3}$$

6. $2f(xy) = \{f(x)\}^y + \{f(y)\}^x$ ... (i)

and $f(1) = k \neq 1$

Substituting $y = 1$ in equation (i), we get

$$2f(x) = \{f(x)\} + \{f(1)\}^x$$

$$\Rightarrow \quad f(x) = \{f(1)\}^x$$

$$\Rightarrow \quad f(x) = k^x$$

$$\therefore \quad \sum_{r=1}^{n} f(r) = \sum_{r=1}^{n} k^r = k + k^2 + k^3 + ... + k^n$$

$$= k\frac{(k^n - 1)}{k - 1} \qquad (\because \text{ G.P. sum formula})$$

7. Given $f(x).f\left(\frac{1}{x}\right) = f(x) + f\left(\frac{1}{x}\right)$

$$\Rightarrow \quad f(x).f\left(\frac{1}{x}\right) - f(x) = f\left(\frac{1}{x}\right)$$

$$\Rightarrow \quad f(x)\left\{f\left(\frac{1}{x}\right) - 1\right\} = f\left(\frac{1}{x}\right)$$

120

$$\Rightarrow \quad f(x) = \frac{f\left(\dfrac{1}{x}\right)}{f\left(\dfrac{1}{x}\right) - 1}$$

$$\Rightarrow \quad f\left(\frac{1}{x}\right) - 1 = \frac{f\left(\dfrac{1}{x}\right)}{f(x)} \qquad \ldots (i)$$

Also, $\quad f(x).f\left(\dfrac{1}{x}\right) - f\left(\dfrac{1}{x}\right) = f(x)$

$$\Rightarrow \quad f(x) - 1 = \frac{f(x)}{f\left(\dfrac{1}{x}\right)} \qquad \ldots (ii)$$

On multiplying (i) and (ii), we get

$$\left[f\left(\frac{1}{x}\right) - 1\right][f(x) - 1] = 1 \qquad \ldots (iii)$$

$\because f(x)$ is a polynomial function.

So, $\{f(x) - 1\}$ *and* $\left[f\left(\dfrac{1}{x}\right) - 1\right]$ are reciprocal of each other, also x

and $\dfrac{1}{x}$ are reciprocal of each other.

Thus (iii) can hold only when

$f(x) - 1 = \pm x^n$ where $n \in R$

$$\Rightarrow \quad f(x) = \pm x^n + 1$$

$\because \quad f(4) = 65$

$$\Rightarrow \quad \pm 4^n + 1 = 65 \Rightarrow n = 3$$

$\therefore \quad f(x) = x^3 + 1$

$\therefore \quad f(6) = 6^3 + 1 = 217$

8.   For all x, $f(f(x)) = x$ *i.e.*,

$$\frac{a\left(\dfrac{ax+b}{cx+d}\right)+b}{c\left(\dfrac{ax+b}{cx+d}\right)+d}=x$$

$$\Rightarrow \quad \frac{(a^2+bc)x+b(a+d)}{c(a+d)x+bc+d^2}=x$$

$$\Rightarrow \quad c(a+d)x^2+(d^2-a^2)x-b(a+d)=0$$

$$\Rightarrow \quad (a+d)\left[cx^2+(d-a)x-b\right]=0$$

$$\Rightarrow \quad a+d=0 \ \ or \ \ cx^2+(d-a)x-b=0$$

$$\Rightarrow \quad a=-d \ \ or \ \ cx^2=(a-d)x+b$$

$$\Rightarrow \quad cx^2=2ax+b$$

$$\therefore \quad f(19)=19 \ \ and \ \ f(97)=97$$

$$\Rightarrow \quad c(19)^2=2a(19)+b \ \ and \ \ c(97)^2=2a(97)+b$$

$$\Rightarrow \quad (97^2+19^2)c=2(97-19)a$$

$$\Rightarrow \quad a=58c \quad and \quad b=-1843\ c$$

$$\therefore \quad f(x)=\frac{58x-1843}{x-58}=58+\frac{1521}{x-58}$$

which never has the value 58

$$\therefore \quad \text{range of } f \text{ is } R\sim\{58\}$$

Alternate Method:

The statement implies that is its own inverse.

$$\therefore \quad y=f(x)=\frac{ax+b}{cx+d}$$

Put $\quad y=x$

$$\therefore \quad x=\frac{ay+b}{cy+d}$$

122

$$\Rightarrow \quad y = f^{-1}(x) = \frac{dx - b}{-cx + a} = \frac{-dx + b}{cx - a}$$

$$\therefore \quad \frac{ax + b}{cx + d} = \frac{b - dx}{cx - a}$$

$$acx^2 - ab - a^2x + bcx$$

$$= bcx + bd - cdx^2 - d^2x$$

$$(a + d)cx^2 + (d^2 - a^2)x - b(a + d) = 0$$

$$\Rightarrow \quad (a + d)\left[cx^2 + (d - a)x - b\right] = 0$$

The solution is same as before.

9. We have

$$f(p, q) = f(p - 1, q) + p - 1$$

$$= f(p - 2, q) + (p - 2) + (p - 1)$$

$$\vdots$$

$$= f(1, q) + \frac{p(p - 1)}{2}$$

$$\vdots$$

$$= f(1, q - 1) - (q - 1) + \frac{p(p - 1)}{2}$$

$$= f(1, 1) - \frac{q(q - 1)}{2} + \frac{p(p - 1)}{2}$$

$$= 2012$$

$$\therefore \quad \frac{p(p - 1)}{2} - \frac{q(q - 1)}{2} = 2011$$

$$\Rightarrow \quad (p - q)(p + q - 1) = 2.(2011)$$

$$\because \quad 2011 \text{ is a prime number and } p - q < p + q - 1 \text{ for } p, q \in N.$$

Case (i) $p - q = 1$ and $p + q - 1 = 4022$

$$\therefore \quad p = 2012 \text{ and } q = 2011$$

Case (ii) $p - q = 2$ and $p + q - 1 = 2011$

$$\therefore \qquad p = 1007 \ and \ q = 1005$$

$$\therefore \qquad (p,\ q) = (2012,\ 2011) \ or \ (1007,\ 1005)$$

10. Case (i) $|a-b| \le \dfrac{1}{2}$

Then, $|f(a)-f(b)| < |a-b| \le \dfrac{1}{2}$

which is proved.

Case (ii) $|a-b| > \dfrac{1}{2}$

By symmetry, we may assume

$$a > b$$

$$\therefore \qquad |f(a)-f(b)| = |f(a)-f(1)+f(0)+f(b)| \quad \left[\because \ f(1)=f(0)=1 \right]$$

$$\Rightarrow \qquad |f(a)-f(b)| |f(a)-f(1)| + |f(0)-f(b)| < |a-1| + |0-b|$$

$$= 1 - a + b - 0$$

$$= 1 - (a-b)$$

$$< \dfrac{1}{2}$$

which is the required result.

11. $f(x) = \dfrac{2}{4^x + 2}$

$$f(1-x) = \dfrac{2}{4^{1-x}+2} = \dfrac{2(4)^x}{4+2(4)^x} = \dfrac{4^x}{4^x+2}$$

$$\therefore \qquad f(x) + f(1-x) = 1$$

[Note: f has a half-turn symmetry about point $\left(\dfrac{1}{2}, \dfrac{1}{2} \right)$].

$$\therefore \qquad Sum = \dfrac{2010}{2} = 1005$$

12. $f(n) = n + f(n-1)$

$$\Rightarrow \qquad f(n) - f(n-1) = n$$

$$f(n-1) - f(n-2) = n-1$$

$$f(n-2) - f(n-1) = n-2$$

$$:$$

$$f(3) - f(2) = 3$$

$$f(2) - f(1) = 2$$

Adding these equations, we get

$$f(n) - f(1) = 2 + 3 + \ldots\ldots + n$$

$$\Rightarrow \quad f(n) = 1 + 2 + 3 + \ldots\ldots + n \qquad \left[\because \quad f(1) = 1\right]$$

$$\Rightarrow \quad f(n) = \frac{n(n+1)}{2}$$

$$\therefore \quad f(2010) = \frac{2010(2011)}{2} = 1005 \times 2011$$

13. $$f(3n) - f(3n-3) = n$$

$$f(3n-3) - f(3n-6) = n-1$$

$$: \quad : \quad :$$

$$f(6) - f(3) = 2$$

Adding these,

$$\therefore \quad f(3n) - f(3) = 2 + 3 + \ldots\ldots + n \qquad \left[\because f(3) = 1\right]$$

$$\Rightarrow \quad f(3n) = 1 + 2 + 3 + \ldots\ldots + n$$

$$\Rightarrow \quad f(3n) = \frac{n(n+1)}{2}$$

$$\therefore \quad f(12) = f(3.4) = \frac{1}{2} 4 \times 5 = 10$$

14. $$f(3n) - f(3n-3) = n^2$$

$$f(3n-3) - f(3n-6) = (n-1)^2$$

$$: \qquad :$$

$$f(6) \quad - \quad f(3) \quad = 2^2$$

Adding these

$$f(3n) - f(3) = 2^2 + 3^2 + \ldots + n^2$$

$$\Rightarrow \quad f(3n) = f(3) + 2^2 + 3^2 + \ldots + n^2$$

$$\Rightarrow \quad f(3n) = 1^2 + 2^2 + 3^2 + \ldots + n^2 \qquad \left[\because f(3) = 1 = 1^2\right]$$

$$\Rightarrow \quad f(3n) = \frac{n(n+1)(2n+1)}{6}$$

$$\therefore \quad f(15) = f(3.5) = \frac{1}{6}(5)(6)(11) = 55$$

15.
$$f(x) = 1 - f(x-1)$$

$$\Rightarrow \quad f(x) + f(x-1) = 1 \qquad \ldots \text{(i)}$$

Put $\quad x = x + 1$

$$\Rightarrow \quad f(x+1) + f(x) = 1 \qquad \ldots \text{(ii)}$$

Subtracting (i) from (ii), we get

$$f(x+1) - f(x-1) = 0 \qquad \Rightarrow \qquad f(x+1) = f(x-1)$$

16. (i) $f = ax + b,\ g = cx + d$

$$\therefore f(g) = a(cx + d) + b = acx + ad + b = x$$

$$\Rightarrow \quad ac = 1,\ ad + b = 0 \ or \ b = -ad$$

(ii) $g(f) = c(ax + b) + d$

$$\Rightarrow \quad g(f) = acx + bc + d$$

$$\because \quad ac = 1 \ and \ b = -ad$$

$$\therefore g(f) = x + (-ad)c + d$$

$$\Rightarrow g(f) = x - (ac)d + d \qquad (\because ac = 1)$$

$$\Rightarrow g(f) = x - d + d$$

$$\Rightarrow g(f) = x$$

$$\therefore \ f(g) = x \Rightarrow g(f)$$

17.
$$f(x) = -x^n (x-1)^n$$

$$f(x^2) = -x^{2n}(x^2 - 1)^n$$

and $\quad f(x+1) = -(x+1)^n (x)^n$

$$\therefore \quad f(x^2) + f(x)f(x+1)$$

$$= -x^{2n}(x^2-1)^n - \left\{(x^n)(x-1)^n\right\}\left\{-(x+1)^n x^n\right\}$$

$$= -x^{2n}(x^2-1)^n + x^{2n}(x^2-1)^n = 0$$

18.   Assume $4f(n) = f(m)$

    $\Rightarrow$ $4n(n+1) = m(m+1)$

    $\Rightarrow$ $4n^2 + 4n = m(m+1)$

    $\Rightarrow$ $4n^2 + 4n + 1 = m^2 + m$

    $\Rightarrow$ $(2n+1)^2 = m^2 + m + 1$

But $m^2 + m + 1$ cannot be the square of an integer.

$\therefore$ There are no natural numbers m and n such that $4f(n) = f(m)$.

Notes: We may say that the product of two successive natural numbers cannot be equal to four times the product of some other pair of successive natural numbers.

EXERCISE - 2   [Time: 3hrs]

1.   Let Z denote the set of all integers. The function $f : Z \to Z$ satisfies the following.

(a) $f(f(n)) = n, \ \forall \ n \in Z$

(b) $f(f(n+2)+2) = n, \ \forall \ n \in Z$

(c) $f(0) = 1$

Find $f(1998)$

2.   The function f defined on the set of ordered pairs of positive integers has the following properties.

(a) $f(x, x) = x$ for all x.

(b) $f(x, y) = f(y, x)$ for all x and y.

(c) $(x+y)f(x, y) = yf(x, x+y)$ for all x and y.

Find $f(52, 14)$.

3.   Find a, b if $g(x) = \dfrac{ax+b}{7x-b}$ satisfies $g(g(x)) = x$ for all $x \neq \dfrac{6}{7}$.

127

Problems in Algebra

4. For any natural number $n(n \geq 3)$, let $f(n)$ denote the number of non-congruent integer sided triangles with perimeter n ($ex: f(3) = 1, f(4) = 0, f(7) = 2$) show that

   (a) $f(1999) > f(1996)$ (b) $f(2000) > f(1997)$

5. A function f defined for all real numbers and for all real x, satisfies the equations

   $f(2+x) = f(2-x)$ *and* $f(7+x) = f(7-x)$.

   If 0 satisfies $f(x) = 0$, what is the smallest number of solutions that $f(x) = 0$ could have in the interval $-2002 \leq x \leq 2002$.

6. Let f be a polynomial of degree 98 such that $f(k) = \dfrac{1}{k}$ for $k = 1, 2, 3, \ldots\ldots 99$. Determine $f(100)$.

7. Find all real valued functions f defined on the set of positive real numbers which satisfy $f(x+y) = f(x^2+y^2)$ for all positive x and y.

8. For any positive integer k. Let $f_1(k)$ denote the square of the sum of the digits of k. For example $f_1(21) = (2+1)^2 = 9$. For $n \geq 2$, let $f_n(k) = f_1(f_{n-1}(k))$. Find $f_{1998}(11)$.

EXERCISE - 2
Answers

1. $f(n) = -n+1;\ f(1998) = -1997$

2. 364

3. $a = 6;\ b \in R \sim \left\{-\dfrac{36}{7}\right\}$

5. 801

6. $\dfrac{1}{50}$

7. Constant function

8. 169

7. INEQUALITIES

1. Elementary Inequalities

    Axiom 1: Trichotomy law

    For any a, b exactly one of the following holds: $a < b$, $a = b$, $a > b$.

    Axiom 2: Transitive law

    If $a < b$ and $b < c$ then $a < c$.

    Axiom 3: Preservation of inequality by addition

    If $a < b$ then $a + c < b + c$ for all $c \in \mathbb{R}$.

    Axiom 4: Preservation of inequality by positive multiplication/ division

    If $a < b$ and $c > 0$, then $ac < bc$ and $\dfrac{a}{c} < \dfrac{b}{c}$.

    Axiom 5: Reversal of inequality by negative multiplication/division.

    If $a < b$ and $c < 0$, then $ac > bc$ and $\dfrac{a}{c} > \dfrac{b}{c}$.

2. Bernoulli's inequality. If $a > 0$ then $(1 + a^n) \geq 1 + na$ for any natural number n.

3. Harder Inequalities

    (i) Inequality : $x + \dfrac{1}{x} \geq 2, \ \forall \ x > 0$.

    (ii) AM-GM-HM inequality

    If a, b are two positive reals then the arithmetic mean A, geometric mean G and harmonic Mean H are defined as follows:

    $$A = \frac{a+b}{2}, \qquad G = \sqrt{ab}, \qquad H = \frac{2ab}{a+b}$$

    If a, b, c are three positive reals then

    $$A = \frac{a+b+c}{3}, \qquad G = (abc)^{\frac{1}{3}}, \qquad H = \frac{3}{\frac{1}{a} + \frac{1}{b} + \frac{1}{c}}.$$

    $$A = \frac{a_1 + a_2 + \dots + a_n}{n}$$

$$G = (a_1, a_2, \ldots\ldots, a_n)^{\frac{1}{n}}$$

$$H = \frac{n}{\frac{1}{a_1} + \frac{1}{a_2} + \ldots\ldots + \frac{1}{an}}$$

These means are related as follows:

$$A \geq G \geq H$$

Equality holds only when all the numbers are equal.

(iii) Weierstrass inequality: If $a_1, a_2, \ldots\ldots, a_n$ are positive real numbers, then, for $n \geq 2$:

$$(1 + a_1)(1 + a_2)\ldots\ldots(1 + a_n) > 1 + (a_1 + a_2 + \ldots\ldots + a_n).$$

If $a_1, a_2, \ldots\ldots, a_n$ are less than unity, then

$$(1 - a_1)(1 - a_2)(1 - a_3)\ldots\ldots(1 - a_n) > 1 - (a_1 + a_2 + \ldots + a_n)$$

(iv) Cauchy-Schwaz inequality: If $(a_1, a_2, \ldots\ldots, a_n)$ and $(b_1, b_2, \ldots\ldots, b_n)$ are two sets of real numbers, then

$$(a_1 b_1 + a_2 b_2 + \ldots\ldots + a_n b_n)^2 \leq \left(a_1^2 + a_2^2 + \ldots\ldots + a_n^2\right)\left(b_1^2 + b_2^2 + \ldots\ldots + b_n^2\right)$$

and the equality holds if and only if $\dfrac{a_1}{b_1} = \dfrac{a_2}{b_2} = \ldots\ldots = \dfrac{a_n}{b_n}$

(v) Tchebychev's inequality

If $a_1, a_2, \ldots\ldots, a_n$, and $b_1, b_2, \ldots\ldots, b_n$ are two sets of real numbers such that

$$a_1 \leq a_2 \leq \ldots\ldots \leq a_n, \qquad b_1 \leq b_2 \ldots\ldots \leq b_n$$

then, $$\left(\frac{a_1 b_1 + a_2 b_2 + a_n b_n}{n}\right) \geq \left(\frac{a_1 + a_2 + \ldots\ldots + a_n}{n}\right)$$

$$\left(\frac{b_1 + b_2 + \ldots\ldots + b_n}{n}\right)$$

(Equality holds when $a_1 = a_2 = \ldots\ldots = a_n$ and $b_1 = b_2 = \ldots\ldots = b_n$.

(vi) Powre mean theorem

If a, b are positive reals then

$$\frac{a^m + b^m}{2} \leq \left(\frac{a+b}{2}\right)^m \qquad if \qquad 0 < m < 1$$

$$\frac{a^m + b^m}{2} \geq \left(\frac{a+b}{2}\right)^m \qquad if \qquad m < 0 \quad or \quad m > 1.$$

The equality holds when $a = b$.

This can be extended to a finite no of reals $a_1, a_2, \ldots\ldots a_n$.

EXERCISE - 1

1. Prove that if $a, b, c, d > 0$ then

$$\frac{cd(a^2 + b^2) + bd(a^2 + c^2)}{abcd} \geq 4$$

2. If a, b and c are real numbers such that $a^2 + b^2 + c^2 = 1$, then what is the maximum value of $ab + bc + ac$?

3. Prove that $a^3 + b^3 + c^3 \geq 3abc$

4. Three real numbers x, y, z such that $x + y + z = 2$ and $x^2 + y^2 + z^2 = 4$ were chosen from the set of all real numbers. Show that each of x, y and z lie in the interval $\left[-\frac{2}{3}, 2\right]$. Are the extreme values attainable?

5. If a and b are positive real numbers such that $a + b = 1$, prove that

$$\left(a + \frac{1}{a}\right)^2 + \left(b + \frac{1}{b}\right)^2 \geq \frac{25}{2}.$$

6. Show that: $n^n \left(\frac{n+1}{2}\right)^{2n} > (n!)^3$

7. If x, y, z be positive real numbers such that $x^2 + y^2 + z^2 = 27$, then show that $x^3 + y^3 + z^3 \geq 81$.

8. If a, b, c are all positive and no two of them are equal, then prove that

(i) $a^3 + b^3 + c^3 > \dfrac{(a+b+c)^3}{9} > 3abc$

(ii) $a^4 + b^4 + c^4 > abc(a+b+c)$

9. If a, b, c be all positive, then show that

$$\frac{a}{b+c} + \frac{b}{c+a} + \frac{c}{a+b} \geq \frac{3}{2}.$$

10. If a, b, c are positive and unequal, show that

$$(a^7 + b^7 + c^7)(a^2 + b^2 + c^2) > (a^5 + b^5 + c^5)(a^4 + b^4 + c^4).$$

11. Show that in any triangle with sides a, b and c, we have

$$(a+b+c)^2 < 4(ab+bc+ca).$$

12. If a, b, c, d be any four positive real numbers, then prove that

$$\frac{a}{b} + \frac{b}{c} + \frac{c}{d} + \frac{d}{a} \geq 4$$

13. If a, b, c, d are four positive real numbers such that $abcd = 1$, prove that

$$(1+a)(1+b)(1+c)(1+d) \geq 16.$$

14. If a, b and c are positive real numbers, show that

$$a^2 b + ab^2 + c^2 a + ca^2 + b^2 c + bc^2 + 2abc \geq 8abc. \cdot$$

15. If a, b, c, d are positive real numbers, then show that

$$(a+b+c+d)\left(\frac{1}{a} + \frac{1}{b} + \frac{1}{c} + \frac{1}{d}\right) \geq 16.$$

What happens when the numbers are all unequal?

16. If a, b, c are positive real numbers, prove that

$$\frac{b^2 + c^2}{b+c} + \frac{c^2 + a^2}{c+a} + \frac{a^2 + b^2}{a+b} \geq a+b+c$$

17. If a, b, c are positive real numbers, prove that

$$6abc \leq a^2(b+c) + b^2(c+a) + c^2(a+b) \leq 2(a^3 + b^3 + c^3)$$

132

18. If a, b, c are any three real numbers, show that

$$a^4 + b^4 + c^2 \geq 2\sqrt{2}\,abc \cdot$$

19. If a, b, c are positive real numbers such that $a + b + c = 1$, prove that

$$(1+a)(1+b)(1+c) \geq 8(1-a)(1-b)(1-c)$$

20. If a, b, $c > 0$ are such that $a + b + c = 1$, prove that $ab + bc + ca \leq \dfrac{1}{3}$.

21. Prove that for each natural number n,

$$\frac{1}{n+1} + \frac{1}{n+2} + \ldots\ldots + \frac{1}{3n+1} > 1.$$

22. Prove that

$$1 < \frac{1}{1001} + \frac{1}{1002} + \frac{1}{1003} + \ldots\ldots + \frac{1}{3001} < 1\frac{1}{3}.$$

23. If a, b are positive real numbers, prove that

$$\left[(a+b)(a^2+b^2)\ldots\ldots(a^n+b^n)\right]^2 > (a^{n+1}+b^{n+1})^n$$

for every positive integer n.

24. Prove that

$$\frac{(a-b)^2}{8a} \leq \frac{a+b}{2} - \sqrt{ab} \leq \frac{(a-b)^2}{8b}$$

for all $a \geq b > 0$

25. Let a, b and c be positive real numbers such that $a + b + c \leq 4$ and $ab + bc + ca \geq 4$.

Prove that at least two of the inequalities

$$|a-b| \leq 2, \qquad |b-c| \leq 2, \qquad |c-a| \leq 2 \text{ are true.}$$

26. Prove that $\dfrac{1}{2} \cdot \dfrac{3}{4} \ldots \ldots \dfrac{2n-1}{2n} < \dfrac{1}{\sqrt{2n}}$ for all positive integers n.

27. Let a, b, c and d be real numbers such that

$$(a^2+b^2-1)(c^2+d^2-1) > (ac+bd-1)^2$$

Prove that $\qquad a^2 + b^2 > 1 \; and \; c^2 + d^2 > 1$

Problems in Algebra

28. Let $P = \left(\dfrac{1}{a}-1\right)\left(\dfrac{1}{b}-1\right)\left(\dfrac{1}{c}-1\right)$ where a, b, c are positive numbers such that $a+b+c=1$. Find the largest integer N such that $P \geq N$.

29. Find the set of values for x such that $x^3 +1 > x^2 + x$.

30. Show that $F = \dfrac{1}{2},\dfrac{3}{4},\dfrac{5}{6},\ldots\dfrac{99}{100} < \dfrac{1}{\sqrt{101}}$

31. Show that $P = \dfrac{2}{3},\dfrac{4}{5},\dfrac{6}{7}\ldots\dfrac{100}{101} > \dfrac{\sqrt{101}}{100}$

32. Which is larger $\sqrt[9]{9!}$ or $\sqrt[10]{10!}$?

33. If x is positive, how large must x be so that $\sqrt{x^2+x}-x$ shall differ from $\dfrac{1}{2}$ by less than 0.02?

34. Find a rational approximation $\dfrac{m}{n}$ to $\sqrt{2}$ such that $-\dfrac{1}{8n} < \sqrt{2}-\dfrac{m}{n} < \dfrac{1}{8n}$ where $n \leq 8$.

35. Find the least value of $(a_1 + a_2 + a_3 + a_4)\left(\dfrac{1}{a_1}+\dfrac{1}{a_2}+\dfrac{1}{a_3}+\dfrac{1}{a_4}\right)$ where each a_i, $i=1,2,3,4$, is positive.

EXERCISE - 1
Solutions

1. $\dfrac{cd(a^2+b^2)+bd(a^2+c^2)}{abcd} = \dfrac{cd(a^2+b^2)}{abcd}+\dfrac{bd(a^2+c^2)}{abcd}$

$= \dfrac{a^2}{ab}+\dfrac{b^2}{ab}+\dfrac{a^2}{ac}+\dfrac{c^2}{ac}$

$= \dfrac{a}{b}+\dfrac{b}{a}+\dfrac{a}{c}+\dfrac{c}{a}$

134

Put $x = \dfrac{a}{b}$ so that $\dfrac{1}{x} = \dfrac{b}{a}$, and $x + \dfrac{1}{x} \geq 2$, observing that $x > 0$ because $a, b > 0$.

$$\Rightarrow \qquad \dfrac{a}{b} + \dfrac{b}{a} \geq 2.$$

Similarly $\dfrac{a}{c} + \dfrac{c}{a} \geq 2;$

$$\therefore \qquad \dfrac{a}{b} + \dfrac{b}{a} + \dfrac{a}{c} + \dfrac{c}{a} \geq 4$$

2. By the Cauchy-Schwarz inequality with $n = 3$ taking (a, b, a) and (b, c, c).

$$(ab + bc + ac)^2 \leq (a^2 + b^2 + a^2)(b^2 + c^2 + c^2)$$

$$= (1 - c^2 + a^2)(1 - a^2 + c^2)$$

$$= (1 + x)(1 - x) \qquad (\text{where } x = a^2 - c^2)$$

$$= 1 - x^2 \leq 1.$$

Now we know that 1 is an upper bound, but is it actually attained?

Clearly it is attained by setting $a = b = c = \sqrt{\dfrac{1}{3}}$.

$$\therefore \qquad ab + bc + ca = 1 \text{ which is the maximum value.}$$

3. The AM-GM inequality gives

$$\dfrac{a^3 + b^3 + c^3}{3} \geq \sqrt[3]{a^3 b^3 c^3}$$

$$\Rightarrow \qquad a^3 + b^3 + c^3 \geq 3abc$$

4. The given system of equations is symmetric in x, y and z, that is, any permutation of the three variables does not change the equation.

This means we need only prove restrictions on x, and they will hold also for y and z. Therefore we aim at an inequality involving just the one variable x:

$$x + y + z = 2 \Rightarrow y + z = 2 - x$$

135

and $\qquad x^2 + y^2 + z^2 = 4 \Rightarrow y^2 + z^2 = 4 - x^2$

For any y and $z \in \mathbb{R}$, we have

$$(y-z)^2 \ge 0 ,$$

$\Rightarrow \qquad\qquad y^2 + z^2 \ge 2zy ,$

hence $\qquad\qquad 2y^2 + 2z^2 \ge y^2 + z^2 + 2zy ,$

$\Rightarrow \qquad\qquad 2(y^2 + z^2) \ge (y+z)^2 ,$

$\Rightarrow \qquad\qquad y^2 + z^2 \ge \dfrac{1}{2}(y+z)^2 , \qquad\qquad .. (1)$

$\because \qquad y^2 + z^2 = 4 - x^2$

and $\qquad y + z = 2 - x$

Substituting in (1), we get

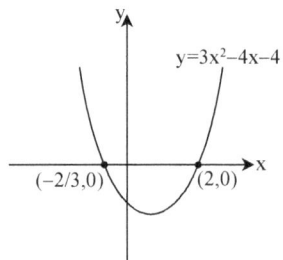

$$4 - x^2 \ge \frac{1}{2}(2-x)^2$$

$\Rightarrow \qquad 3x^2 - 4x - 4 \le 0$ which is shown in graph also.

$\therefore \qquad (3x+2)(x-2) \le 0, \ or \ x \in \left[-\dfrac{2}{3}, 2 \right]$

To see whether extreme values are attainable or not, try $x = 2$, giving $y + z = 0$ and $y^2 + z^2 = 4$, which is obviously possible, as $z = 0$, $y = 2$ and $z = 2$, $y = 0$ are solutions. next, trying $x = -\dfrac{2}{3}$ gives

136

$$y + z = \frac{8}{3}, \text{ and } y^2 + z^2 = \frac{32}{9},$$

therefore $\qquad \left(\frac{8}{3} - z\right)^2 + z^2 = \frac{32}{9},$

so that $\qquad 2z^2 - \frac{16z}{3} + \frac{32}{9} = 0,$

and hence $\qquad 18z^2 - 48z + 32 = 0,$

Now the discriminant of this equation is: $\Delta = b^2 - 4ac = (-48)^2 - 4.18.32 = 0,$ so there is a real root for z (and the same for y by symmetry). Therefore $x = -\frac{2}{3}$ is attainable. hence x (and by symmetry y and z too) can take either of the extreme values 2 and $-\frac{2}{3}$.

5. $\quad \left(a + \frac{1}{a}\right)^2 + \left(b + \frac{1}{b}\right)^2 = a^2 + \frac{1}{a^2} + b^2 + \frac{1}{b^2} + 4$

$$= (a+b)^2 - 2ab + \left(\frac{1}{a} + \frac{1}{b}\right)^2 - \frac{2}{ab} + 4$$

$$= 1 - 2ab + \frac{1 - 2ab}{a^2 b^2} + 4 \qquad \qquad \dots (1)$$

$\because \qquad (a-b)^2 \geq 0,$

therefore, $\qquad a^2 - 2ab + b^2 \geq 0,$

$\Rightarrow \qquad (a+b)^2 - 4ab \geq 0$

$\Rightarrow \qquad (a+b)^2 \geq 4ab,$

giving $\qquad \left(\frac{a+b}{2}\right)^2 \geq ab, \qquad \qquad (2)$

hence $\qquad ab \leq \frac{1}{4} \qquad \qquad (3)$

Problems in Algebra

$$ab \leq \frac{1}{4} \Rightarrow 2ab \leq \frac{1}{2} \Rightarrow -2ab \geq -\frac{1}{2};$$

$$ab \leq \frac{1}{4} \Rightarrow \frac{1}{ab} \geq 4 \Rightarrow \frac{1}{a^2b^2} \geq 16$$

Finally, using these inequalities in equation (1) gives

$$\left(a + \frac{1}{a}\right)^2 + \left(b + \frac{1}{b}\right)^2 \geq \left(1 - \frac{1}{2}\right) + 16\left(1 - \frac{1}{2}\right) + 4 = \frac{25}{2}.$$

6.   Consider the unequal positive numbers

$$1^3, 2^3, \dots\dots\dots, n^3$$

Applying A.M. > G.M.,

$$\therefore \qquad \frac{1^3 + 2^3 + \dots\dots\dots + n^3}{n} > (1^3.2^3\dots..n^3)^{1/n}$$

$$\Rightarrow \qquad \frac{n(n+1)^2}{4} > \left\{(n!)^3\right\}^{1/n}$$

$$\Rightarrow \qquad n^n \left(\frac{n+1}{2}\right)^{2n} > (n!)^3$$

7.   Applying Cauchy-Schwarz inequality to the two sets of numbers

$$x^{3/2}, y^{3/2}, z^{3/2} \; ; x^{1/2}, y^{1/2}, z^{1/2}, \text{ we get}$$

$$(x^2 + y^2 + z^2)^2 \leq (x^3 + y^3 + z^3)(x + y + z) \qquad \dots \text{(i)}$$

Again, applying Cauchy-Schwarz inequality to the two sets of numbers

$$x, y, z \; ; 1, 1, 1, \text{ we get}$$

We have $(x + y + z)^2 \leq 3(x^2 + y^2 + z^2)$ \qquad ... (ii)

Squaring both sides of (i), we get

$$(x^2 + y^2 + z^2)^4 \leq (x^3 + y^3 + z^3)^2 (x + y + z)^2$$

On using (ii), the above inequality yields,

$$(x^2 + y^2 + z^2)^4 \leq 3(x^3 + y^3 + z^3)^2 (x^2 + y^2 + z^2) \qquad \dots \text{(iii)}$$

$$\therefore \qquad x^2 + y^2 + z^2 = 27, \text{ we have from (iii)}$$

138

$$(x^3 + y^3 + z^3)^2 \geq (81)^2.$$

Taking positive square roots, we get

$$x^3 + y^3 + z^3 \geq 81.$$

8. (i) Assume that $a < b < c$. By applying the generalised Tchebychef's inequality to three sets of numbers each of which is the same as a, b, c, we obtain

$$\frac{a^3 + b^3 + c^3}{3} > \frac{a+b+c}{3} \cdot \frac{a+b+c}{3} \cdot \frac{a+b+c}{3}$$

i.e., $\qquad a^3 + b^3 + c^3 > (a+b+c)^3 / 9 \qquad \qquad$ ... (1)

Again, since the arithmetic mean exceeds the geometric mean

$$\left(\frac{a+b+c}{3}\right)^3 > abc \qquad \qquad \text{... (2)}$$

From (1) and (2), we obtain the inequalities

$$a^3 + b^3 + c^3 > (a+b+c)^3 / 9 > 3abc \qquad \qquad \text{... (} i \text{)}$$

(ii) Assume that $a < b < c$.

Since $a < b < c$, therefore, $a^3 < b^3 < c^3$.

Applying Tchebychef's inequality to the sets of numbers a, b, c ; a^3, b^3, c^3 , we obtain

$$\frac{a^4 + b^4 + c^4}{3} > \frac{a^3 + b^3 + c^3}{3} \cdot \frac{a+b+c}{3} \qquad \qquad \text{... (3)}$$

Also, from (i) , $\quad \dfrac{a^3 + b^3 + c^3}{3} > abc$, $\qquad \qquad$ ... (4)

From (3) and (4), we have

$$a^4 + b^4 + c^4 > abc(a+b+c).$$

Remarks 1. The strict inequalities in (1), (2) and (3) holds because no two of the numbers a, b, c are equal.

2. Since the inequalities are symmetric in a, b, c therefore no generality is lost by assuming $a < b < c$.

9. Since a, b, c are positive, therefore, $b+c, c+a, a+b$ are positive

$$\therefore \qquad \frac{1}{b+c}, \frac{1}{c+a}, \frac{1}{a+b} \text{ are also positive.}$$

Let us assume that $b+c \le c+a \le a+b$.

Then $\dfrac{1}{b+c} \ge \dfrac{1}{c+a} \ge \dfrac{1}{a+b}$.

Applying Tchebychef's inequality to the sets of numbers

$b+c, c+a, a+b ; \dfrac{1}{b+c}, \dfrac{1}{c+a}, \dfrac{1}{a+b}$, we obtain

$$\{(b+c)+(c+a)+(a+b)\} \left\{ \frac{1}{b+c} + \frac{1}{c+a} + \frac{1}{a+b} \right\}$$

$$\ge 3 \left\{ (b+c)\frac{1}{b+c} + (c+a)\frac{1}{c+a} + (a+b)\frac{1}{a+b} \right\},$$

$$\Rightarrow \qquad \frac{a+b+c}{b+c} + \frac{a+b+c}{c+a} + \frac{a+b+c}{a+b} \ge \frac{9}{2}.$$

$$\Rightarrow \qquad \frac{a}{b+c} + \frac{b}{c+a} + \frac{c}{a+b} \ge \frac{3}{2}.$$

10. $(a^7 + b^7 + c^7)(a^2 + b^2 + c^2) - (a^5 + b^5 + c^5)(a^4 + b^4 + c^4)$

$$= \Sigma(a^7 b^2 + a^2 b^7 - a^5 b^4 - a^4 b^5)$$

$$= \Sigma a^2 b^2 (a^5 + b^5 - a^3 b^2 - a^2 b^3)$$

$$= \Sigma a^2 b^2 (a^3 - b^3)(a^2 - b^2)$$

The differences $a^2 - b^2, a^3 - b^3$ are both of the same sign, and therefore, $(a^2 - b^2)(a^3 - b^3)$ is positive, Similarly, the other two terms in the above sum are also positive. Therefore,

$$(a^7 + b^7 + c^7)(a^2 + b^2 + c^2) - (a^5 + b^5 + c^5)(a^4 + b^4 + c^4) > 0$$

11. Let us assume that $a > b > c > 0$. Since the sum of any two sides of a triangle is always greater than the third side, therefore the difference between any two sides can never exceed the third, and consequently

$$0 \le b - c < a, \ 0 \le a - c < b, \ 0 < a - b < c.$$

Squaring both sides of the above inequalities and adding the corresponding sides, we have

$$(b-c)^2 + (a-c)^2 + (a-b)^2 < a^2 + b^2 + c^2$$

$$\Leftrightarrow \qquad\qquad a^2 + b^2 + c^2 < 2(bc + ca + ab)$$

$$\Leftrightarrow \qquad\qquad (a+b+c)^2 < 4(bc + ca + ab)$$

12. Applying the inequality of the mean (AM ≥ GM) to the four positive numbers $\dfrac{a}{b}, \dfrac{b}{c}, \dfrac{c}{d}, \dfrac{d}{a}$, we have

$$\frac{1}{4}\left(\frac{a}{b} + \frac{b}{c} + \frac{c}{d} + \frac{d}{a}\right) \geq \left(\frac{a}{b} \times \frac{b}{c} \times \frac{c}{d} \times \frac{d}{a}\right)^{\frac{1}{4}},$$

i.e., $\qquad \dfrac{a}{b} + \dfrac{b}{c} + \dfrac{c}{d} + \dfrac{d}{a} \geq 4$.

13. Since AM ≥ GM, therefore

$$\frac{1+a}{2} \geq \sqrt{(1.a)}, \frac{1+b}{2} \geq \sqrt{(1.b)} , \frac{1+c}{2} \geq \sqrt{(1.c)}, \frac{1+d}{2} \geq \sqrt{(1.d)}$$

Multiplying corresponding sides of the above inequalities, we have

$$(1+a)(1+b)(1+c)(1+d) \geq 16\sqrt{(abcd)} , \qquad \left(\because \ abcd = 1\right)$$

$$\therefore \ (1+a)(1+b)(1+c)(1+d) \geq 16.$$

14. $a^2b + ab^2 + c^2a + ca^2 + b^2c + bc^2 + 2abc$

$$= (b+c)(c+a)(a+b) ,$$

$$\geq 2(bc)^{1/2}.2(ca)^{1/2}.2(ab)^{1/2} = 8abc$$

Aliternate Method:

$$a^2b + ab^2 + c^2a + ca^2 + b^2c + bc^2 + 3abc$$

$$= (a+b+c)(bc + ca + ab)$$

$$\geq 3(abc)^{1/3}.3(bc.ca.ab)^{1/3} ,$$

$$= 9abc \cdot$$

15. By applying the inequality A.M. ≥ G.M. successively to the sets of the numbers a, b, c, d; $1/a, 1/b, 1/c, 1/d$, we have

$$\frac{a+b+c+d}{4} \geq (abcd)^{\frac{1}{4}} \qquad \ldots (1)$$

and $\left(\dfrac{1}{a}+\dfrac{1}{b}+\dfrac{1}{c}+\dfrac{1}{d}\right)\Big/ 4 \geq \left(\dfrac{1}{a}\cdot\dfrac{1}{b}\cdot\dfrac{1}{c}\cdot\dfrac{1}{d}\right)^{\frac{1}{4}} \qquad \ldots (2)$

By multiplying corresponding sides of the above inequalities we

get $(a+b+c+d)\left(\dfrac{1}{a}+\dfrac{1}{b}+\dfrac{1}{c}+\dfrac{1}{d}\right) \geq 16$.

Since the equality holds in A.M. $\geq$ G.M. if all the number are equal, therefore if the numbers a, b, c, d are not all equal, the strict inequality holds. The condition a, b, c, d are all unequal is a stronger condition than the condition the numbers are not all equal. Therefore when the numbers are all unequal, the strict inequality holds.

16. $b^2 + c^2 \geq \dfrac{1}{2}(b+c)^2 \qquad \left[\because (b+c)^2 + (b-c)^2 = 2(b^2+c^2)\right]$

$\Rightarrow \qquad \dfrac{b^2+c^2}{b+c} \geq \dfrac{1}{2}(b+c)$

Similarly, $\dfrac{c^2+a^2}{c+a} \geq \dfrac{1}{2}(c+a)$

$\dfrac{a^2+b^2}{a+b} \geq \dfrac{1}{2}(a+b)$

Adding corresponding sides of the above inequalities, we obtain

$$\frac{b^2+c^2}{b+c} + \frac{c^2+a^2}{c+a} + \frac{a^2+b^2}{a+b} \geq a+b+c.$$

17. By the inequality of the means,

$$\frac{b+c}{2} \geq (bc)^{\frac{1}{2}}, \frac{c+a}{2} \geq (ca)^{\frac{1}{2}}, \frac{a+b}{2} \geq (ab)^{\frac{1}{2}}$$

so that $a^2(b+c) + b^2(c+a) + a^2(b+c)$

$$\geq 2\left\{ a^2(bc)^{\frac{1}{2}}.b^2(ca)^{\frac{1}{2}}.c^2(ab)^{\frac{1}{2}} \right\} \qquad \ldots (1)$$

Also, by the inequality of the means,

$$\frac{a^2(bc)^{\frac{1}{2}}+b^2(ca)^{\frac{1}{2}}+c^2(ab)^{\frac{1}{2}}}{3} \geq \left\{a^2(bc)^{\frac{1}{2}}.b^2(ca)^{\frac{1}{2}}.c^2(ab)^{\frac{1}{2}}\right\}^{\frac{1}{3}}$$

$$= abc \qquad \qquad \dots(2)$$

Find (1) and (2), we find that

$$a^2(b+c)+b^2(c+a)+c^2(a+b) \geq 6abc \qquad \dots(3)$$

Since $a-b$ and a^2-b^2 are both of the same sign (either $a-b>0$ and $a^2-b^2>0$, or $0<a<b$ and $0<a^2<b^2$). In either case $(a-b)(a^2-b^2)>0$, so that $a^3+b^3>a^2b+ab^2$. Similarly,

$$b^3+c^3>b^2c+bc^2,\ c^3+a^3>c^2a+ca^2.$$

Adding corresponding sides, we have

$$2(a^3+b^3+c^3)>a^2(b+c)+b^2(c+a)+c^2(a+b)$$

18. Let a, b be any real numbers. Therefore by applying the inequality of the means to the non-negative real numbers a^4, b^4 we have

$$a^4+b^4 \geq 2a^2b^2,$$

$$\Rightarrow \qquad a^4+b^4+c^2 \geq 2a^2b^2+c^2$$

$$\geq 2\sqrt{(2a^2b^2.c^2)} \qquad \qquad \dots(1)$$

From (1), we have

$$a^4+b^4+c^2 \geq 2\sqrt{2}\ |abc| \geq 2\sqrt{2}\ abc.$$

19. $(1+a)(1+b)(1+c) \geq 8(1-a)(1-b)(1-c)$ $\qquad (\because 1=a+b+c)$

Substituting, we get equivalent inequality,

$$(2a+b+c)(a+2b+c)(a+b+2c) \geq 8(b+c)(c+a)(a+b).$$

Let $\dfrac{b+c}{2}=x,\ \dfrac{c+a}{2}=y,\ \dfrac{a+b}{2}=z$,

then $x+y+z = \dfrac{b+c}{2}+\dfrac{c+a}{2}+\dfrac{a+b}{2} = a+b+c = 1,\ x,y,z>0$

and the given inequality becomes

$$(y+z)(z+x)(x+y) \geq 8 \; xyz \qquad \qquad \dots (1)$$

In order to prove the given inequality, it is enough to prove (1).

Since x, y, z are all positive, applying inequality A.M > G.M

$$\frac{y+z}{2} \geq (yz)^{\frac{1}{2}}, \; \frac{z+x}{2} \geq (zx)^{\frac{1}{2}}, \; \frac{x+y}{2} \geq (xy)^{\frac{1}{2}},$$

Now $(y+z)(z+x)(x+y)$

$$\geq \left[2(yz)^{\frac{1}{2}} \right] . \left[2(zx)^{\frac{1}{2}} \right] . \left[2(xy)^{\frac{1}{2}} \right]$$

$$= 8 \; xyz \; .$$

which proves (1).

20.   For all real numbers a, b, c

$$(b-c)^2 + (c-a)^2 + (a-b)^2 \geq 0$$

so that $a^2 + b^2 + c^2 \geq ab + bc + ca$

$\Leftrightarrow$ $(a+b+c)^2 \geq 3(ab + bc + ca)$

Since $a+b+c = 1$, we get

$$1 \geq 3(ab + bc + ca)$$

i.e., $ab + bc + ca \leq \dfrac{1}{3}$.

21. $\dfrac{1}{n+1} + \dfrac{1}{n+2} + \dfrac{1}{n+3} + \dots\dots + \dfrac{1}{3n+1}$

$$= \left(\frac{1}{n+1} + \frac{1}{3n+1} \right) + \left(\frac{1}{n+2} + \frac{1}{3n} \right) + \dots\dots + \left(\frac{1}{2n} + \frac{1}{2n+2} \right) + \left(\frac{1}{2n+1} \right).$$

Since the A.M. of two unequal positive number exceeds their H.M., therefore

$$\frac{1}{2} \left(\frac{1}{n+1} + \frac{1}{3n+1} \right) > \left\{ \frac{(n+1)+(3n+1)}{2} \right\}^{-1} = \frac{1}{2n+1},$$

$$\frac{1}{2} \left(\frac{1}{n+2} + \frac{1}{3n} \right) > \left\{ \frac{(n+2)+3n}{2} \right\}^{-1} = \frac{1}{2n+1}$$

$$\frac{1}{2}\left(\frac{1}{2n}+\frac{1}{2n+1}\right) > \left\{\frac{2n+(2n+2)}{2}\right\}^{-1} = \frac{1}{2n+1}$$

Adding corresponding sides of the inequalities, we have

$$\frac{1}{2}\left\{\frac{1}{n+1}+\frac{1}{n+2}+.........+\frac{1}{2n}+\frac{1}{2n+2}+......+\frac{1}{3n+1}\right\} > \frac{n}{2n+1}$$

Multiplying throughout by 2 and adding $\dfrac{1}{2n+1}$ to both sides, we get the result.

22. As the arithmetic mean of any n distinct positive numbers always exceeds their harmonic mean. We shall apply this result to the 2001 numbers.

$$1001, 1002, ..., 3000, 3001.$$

For these numbers,

$$A.M. = \frac{1001+1002+.......+3001}{2001}$$

$$= \frac{2001}{2}(1001+3001).\frac{1}{2001}$$

$$= 2001$$

$$H.M. = \left\{\frac{\dfrac{1}{1001}+\dfrac{1}{1002}+......+\dfrac{1}{3000}+\dfrac{1}{3001}}{2001}\right\}^{-1}$$

$$= 2001\Big/\left\{\frac{1}{1001}+\frac{1}{1002}+......+\frac{1}{3001}\right\}$$

Since A.M. > H.M., therefore it follows from the above that

$$2001 > 2001\Big/\left\{\frac{1}{1001}+\frac{1}{1002}+......+\frac{1}{3001}\right\}$$

i.e., $$\frac{1}{1001}+\frac{1}{1002}+.......+\frac{1}{3001} > 1,$$

Also, we have
$$\frac{1}{1001}+\frac{1}{1002}+\ldots\frac{1}{1250}<\frac{250}{1000}=\frac{1}{4}$$

$$\frac{1}{1251}+\frac{1}{1252}+\ldots\ldots+\frac{1}{1500}<\frac{250}{1500}=\frac{1}{5}$$

$$\frac{1}{1501}\ldots\ldots+\frac{1}{2000}<\frac{500}{1500}=\frac{1}{3}$$

$$\frac{1}{2001}+\frac{1}{2002}+\ldots\ldots+\frac{1}{3001}<\frac{1001}{2000}$$

Adding throughout, we have

$$\frac{1}{1001}+\frac{1}{1002}+\ldots\ldots+\frac{1}{3001}<\frac{1}{4}+\frac{1}{5}+\frac{1}{3}+\frac{1001}{2000}$$

$$=\frac{7703}{6000}<1.\frac{1}{3}.$$

23. $$\left[(a+b)(a^2+b^2)\ldots\ldots(a^n+b^n)\right]^2$$

$$=\left[(a+b)(a^2+b^2)\ldots..(a^n+b^n)\right]\left[(a^n+b^n)(a^{n-1}+b^{n-1})\ldots\ldots(a+b)\right],$$

$$=\left[(a+b)(a^n+b^n)\right]\left[(a^2+b^2)(a^{n-1}+b^{n-1})\right]\ldots..\left[(a^n+b^n)(a+b)\right]$$

If k be any positive integer such that $1\le k\le n$, then

$$(a^k+b^k)(a^{n-k+1}+b^{n-k+1})$$

$$=a^{n+1}+b^{n+1}+a^k b^{n-k+1}+b^k a^{n-k+1}>a^{n+1}+b^{n+1}$$

Applying this result to each of the n products enclosed in the brackets, we find that

$$\left[(a+b)(a^2+b^2)\ldots\ldots(a^n+b^n)^2\right]^2$$

$$>\left(a^{n+1}+b^{n+1}\right)\left(a^{n+1}+b^{n+1}\right)\ldots\text{ to }n\text{ factors.}$$

$$=\left(a^{n+1}+b^{n+1}\right)^n$$

Hence the result.

24. Note that $\left(\dfrac{\sqrt{a}+\sqrt{b}}{2\sqrt{a}}\right)^2 \le 1 \le \left(\dfrac{\sqrt{a}+\sqrt{b}}{2\sqrt{b}}\right)^2$

    i.e., $\dfrac{\left(\sqrt{a}+\sqrt{b}\right)^2\left(\sqrt{a}-\sqrt{b}\right)^2}{4a} \le \left(\sqrt{a}-\sqrt{b}\right)^2 \le \dfrac{\left(\sqrt{a}+\sqrt{b}\right)^2\left(\sqrt{a}-\sqrt{b}\right)^2}{4b}$

    i.e., $\dfrac{(a-b)^2}{8a} \le \dfrac{a-2\sqrt{ab}+b}{2} \le \dfrac{(a-b)^2}{8b}$

    from which the result follows.

25. We have, $(a+b+c)^2 \le 16$,

    i.e., $a^2+b^2+c^2+2(ab+bc+ca) \le 16,$

    i.e., $a^2+b^2+c^2 \le 8$

    i.e., $a^2+b^2+c^2-(ab+bc+ca) \le 4$

    i.e., $(a-b)^2+(b-c)^2+(c-a)^2 \le 8$

    and the desired result follows.

26. We prove a stronger statement:

    $$\frac{1}{2}\cdot\frac{3}{4}.....\frac{2n-1}{2n} \le \frac{1}{\sqrt{3n+1}}$$

    We use induction.

    For $n=1$, the result is evident.

    Suppose the statement is true for some positive integer k, i.e.,

    $$\frac{1}{2}\cdot\frac{3}{4}.....\frac{2k-1}{2k} < \frac{1}{\sqrt{3k+1}}$$

    Then, $\dfrac{1}{2}\cdot\dfrac{3}{4}.....\dfrac{2k-1}{2k}\cdot\dfrac{2k+1}{2k+2} < \dfrac{1}{\sqrt{3k+1}}\cdot\dfrac{2k+1}{2k+2}$

    In order for the induction step to pass it suffices to prove that

    $$\frac{1}{\sqrt{3k+1}}\cdot\frac{2k+1}{2k+2} < \frac{1}{\sqrt{3k+4}}$$

    This reduces to

$$\left(\frac{2k+1}{2k+2}\right)^2 < \frac{3k+1}{3k+4}$$

i.e., $\quad (4k^2+4k+1)(3k+4) < (4k^2+8k+4)(3k+1)$

i.e., $\qquad\qquad\qquad 0 < k,$

which is evident. Our proof is complete.

Comment: By using Stirling number, the upper bound can be improved to $1/\sqrt{\pi n}$ sufficiently large n.

27. For the sake of the contradiction, suppose that one of a^2+b^2 or c^2+d^2 is less than or equal to 1. Since $(ac+bd-1)^2 \geq 0$, a^2+b^2-1 and c^2+d^2-1 must have the same sign. Thus both a^2+b^2 and c^2+d^2 are less than 1. Let

$$x = 1-a^2-b^2 \quad and \quad y = 1-c^2-d^2$$

Then $0 < x, y \leq 1$. Multiplying by 4 on both sides of the given inequality gives,

$$4xy > (2ac+2bd-2)^2 = (2-2ac-2bd)^2$$

$$= (a^2+b^2+x+c^2+d^2+y-2ac-2bd)^2$$

$$= \left[(a-c)^2+(b-d)^2+x+y\right]^2$$

$$\geq (x+y)^2 = x^2+2xy+y^2$$

or $0 > x^2-2xy+y^2 = (x-y)^2$, which is impossible.

Thus our assumption is wrong and both a^2+b^2 and c^2+d^2 are greater than 1.

28. $\quad P = \left(\frac{1}{a}-1\right)\left(\frac{1}{b}-1\right)\left(\frac{1}{c}-1\right)$

$$P = \frac{1}{abc} - \left(\frac{1}{ac}+\frac{1}{bc}+\frac{1}{ab}\right) + \left(\frac{1}{a}+\frac{1}{b}+\frac{1}{c}\right) - 1$$

Since $\dfrac{1}{ac}+\dfrac{1}{bc}+\dfrac{1}{ab} = \dfrac{b+a+c}{abc}$ and $a+b+c=1$.

$$P = \frac{1}{abc} - \frac{1}{abc} + \left(\frac{1}{a}+\frac{1}{b}+\frac{1}{c}\right) - 1$$

Therefore, $P = \dfrac{1}{a} + \dfrac{1}{b} + \dfrac{1}{c} - 1$

Since $a + b + c = 1$, $(a + b + c)P = P$

$= (a + b + c)\left(\dfrac{1}{a} + \dfrac{1}{b} + \dfrac{1}{c}\right) - 1$

But $(a + b + c)\left(\dfrac{1}{a} + \dfrac{1}{b} + \dfrac{1}{c}\right) \geq 3^2$.

Therefore, $P \geq 9 - 1 = 8$.

29. Algebraically

$(x^3 + 1 > x^2 + x) \Rightarrow \left[(x + 1)(x^2 - x + 1) > x(x + 1)\right]$

Therefore, when $x + 1 > 0$, $x^2 - x + 1 > x$

But $(x^2 - x + 1 > x) \Rightarrow \left[(x - 1)^2 > 0\right]$

Therefore, $x \neq 1$.

Since $x + 1 > 0$, $x > -1$. Therefore, $x^3 + 1 > x^2 + x$ for $x > -1$, except for $x = 1$.

Stated otherwise, $x^3 + 1 > x^2 + x$ when $-1 < x < 1$, or when $x > 1$.

30. Let $G = \dfrac{2}{3}, \dfrac{4}{5}, \dfrac{6}{7}, \ldots \dfrac{100}{101}$

Since $\dfrac{1}{2} < \dfrac{2}{3}, \dfrac{3}{4} < \dfrac{4}{5}, \ldots, \dfrac{k}{k+1} < \dfrac{k+1}{k+2}, \ldots, F < G$.

$\therefore \quad F^2 < FG = \dfrac{1}{2}, \dfrac{2}{3}, \dfrac{3}{4} \ldots \dfrac{99}{100} \cdot \dfrac{100}{101}$

$\Rightarrow \quad F^2 < \dfrac{1}{101}, \; F < \dfrac{1}{\sqrt{101}}$

31. Since $\dfrac{1}{2} < \dfrac{2}{3}, \dfrac{3}{4} < \dfrac{4}{5}, \ldots, F < P$

But $FP = \dfrac{1}{101}$,

$$\therefore \qquad P > \frac{1}{101} \div \frac{1}{\sqrt{101}} \text{ and } P > \frac{\sqrt{101}}{101}.$$

32. We prove that the positive geometric mean $G_{n+1} = \sqrt[n+1]{(n+1)!}$ is greater than $G_n = \sqrt[n]{n!}$, and hence, $\sqrt[10]{10!} > \sqrt[9]{9!}$.

Since $\sqrt[n+1]{n+1} > \sqrt[n+1]{n}$, $\sqrt[n+1]{n!(n+1)} > \sqrt[n+1]{n!n}$ (multiplying by $\sqrt[n+1]{n!}$); that is, $\sqrt[n+1]{(n+1)!} > \sqrt[n+1]{n!n}$.

Since $n^n > n!$ for $n > 1$, $n > \sqrt[n]{n!}$.

$\therefore n!n > n!(n!)^{\frac{1}{n}} = (n!)^{\frac{n+1}{n}}$, and $\sqrt[n+1]{n!n} > \sqrt[n]{n!}$.

Thus, $\sqrt[n+1]{(n+1)!} > \sqrt[n+1]{n!n} > \sqrt[n]{n!}$.

33. $\left| \sqrt{x^2 + x} - x - \frac{1}{2} \right| < \in$, so we replace .02 by $\in$ and, hence, solve a more general problem.

$$\frac{1}{2} - \in < \sqrt{x^2 + x} - x < \frac{1}{2} + \in.$$

But, $\sqrt{x^2 + x} - x = \dfrac{x}{\sqrt{x^2 + x} + x} = \dfrac{1}{\sqrt{1 + \dfrac{1}{x}} + 1}$

$\therefore \qquad \dfrac{1}{\sqrt{1 + \dfrac{1}{x}} + 1} > \dfrac{1}{2} - \in = \dfrac{1 - 2 \in}{2}$, and $\sqrt{1 + \dfrac{1}{x}} + 1 < \dfrac{2}{1 - 2 \in}$;

$\sqrt{1 + \dfrac{1}{x}} < \dfrac{1 + 2 \in}{1 - 2 \in}$; $1 + \dfrac{1}{x} < \dfrac{(1 + 2 \in)^2}{(1 - 2 \in)^2}$;

$\dfrac{1}{x} < \dfrac{8 \in}{(1 - 2 \in)^2}$

$\therefore \qquad x > \dfrac{(1 - 2 \in)^2}{8 \in}$

For *for* $\in = .02$, $x > 5.76$.

34. Since $-\dfrac{1}{8n} < \sqrt{2} - \dfrac{m}{n} < \dfrac{1}{8n}$, $-\dfrac{1}{8} < n\sqrt{2} - m < \dfrac{1}{8}$. Since $\dfrac{1}{8} = .125$, we list below, for $n \le 8$, the approximate values (three decimal places) of $n\sqrt{2} - m$ such that $n\sqrt{2} - m < 1$.

$$\sqrt{2} - 1 = .414 \qquad\qquad 5\sqrt{2} - 7 = .070$$
$$2\sqrt{2} - 2 = .828 \qquad\qquad 6\sqrt{2} - 8 = .484$$
$$3\sqrt{2} - 4 = .242 \qquad\qquad 7\sqrt{2} - 9 = .898$$
$$4\sqrt{2} - 5 = .656 \qquad\qquad 8\sqrt{2} - 11 = .312$$

From the table we find $-\dfrac{1}{8} < 5\sqrt{2} - 7 < \dfrac{1}{8}$. Therefore,

$-\dfrac{1}{8.5} < \sqrt{2} - \dfrac{7}{5}, \dfrac{1}{8.5}$ so that the required $\dfrac{m}{n}$ is $\dfrac{7}{5}$.

35. We prove more generally that

$$(a_1 + a_2 + \ldots + a_n)\left(\dfrac{1}{a_1} + \dfrac{1}{a_2} + \ldots + \dfrac{1}{a_n}\right) \ge n^2$$

where each a_i, $i = 1, 2, \ldots, n$, is positive so that the answer to the given problem is $4^2 = 16$.

By definition, the harmonic mean (H.M.,) of positive number is

$$H.M. = \left(\dfrac{a_1^{-1} + a_2^{-1} + \ldots + a_n^{-1}}{n}\right)^{-1} = \dfrac{n}{\dfrac{1}{a_1} + \dfrac{1}{a_2} + \ldots + \dfrac{1}{a_n}}.$$ Since the H.M.

$\le$ A.M. (arithmetic mean), then $\dfrac{n}{\dfrac{1}{a_1} + \dfrac{1}{a_2} + \ldots + \dfrac{1}{a_n}} \le \dfrac{a_1 + a_2 + \ldots + a_n}{n}$.

Therefore, $(a_1 + a_2 + \ldots + a_n)\left(\dfrac{1}{a_1} + \dfrac{1}{a_2} + \ldots + \dfrac{1}{a_n}\right) \ge n^2$

151

EXERCISE - 2   [Time: 3hrs]

1. If a, b, c, x, y, z are all positive, prove that $(x+y+z)\left(\dfrac{1}{x}+\dfrac{1}{y}+\dfrac{1}{z}\right)\geq 9$

2. If a, b, c be real numbers such that $a+b+c=1$.

    Prove that $a^2+b^2+c^2 \geq 4(ab+bc+ca)-1$

    When does the equality hold?

3. If a, b, c are the sides of a triangle, prove

    $$\frac{a}{c+a-b}+\frac{b}{a+b-c}+\frac{c}{b+c-a}\geq 3$$

4. If x, y are positive real numbers such that $x+y=1$ prove that

    $\left(1+\dfrac{1}{x}\right)\left(1+\dfrac{1}{y}\right)\geq 9$.

5. Let a, b, c be nonnegative reals such that $a+b+c=1$. Prove that
    $a^2+b^2+c^2+\sqrt{12abc}\leq 1$.

6. Find all solutions in positive real numbers a, b, c, d to the following system, of equations

    $$a+b+c+d=12, \text{ and}$$

    $$abcd = 27 + ab + ac + ad + bc + bd + cd$$

7. For every positive integer 'n' prove that

    $$\sqrt{4n+1} < \sqrt{n}+\sqrt{n+1} < \sqrt{4n+2}$$

8. Prove that if 'n' is a positive integer such that $n \geq 4011^2$, then there exists an integer l such that $n < l^2 < \left(1+\dfrac{1}{2005}\right)n$.

9. Find the smallest positive integer M for which whenever an integer n is such that $n \geq M$, there exists an integers l, such that

    $n < l^2 < \left(1+\dfrac{1}{2005}\right)n$.

EXERCISE - 2
Answers

6. a = b = c = d = 3

8. LOGARITHMS

1. Definition:

   Let 'a' be a real number such that $a > 0$ and $a \neq 1$ and let $a^x = y$, then x is called the logarithms of y to the base 'a' and is written as $\log_a y$. It is obvious that 'y' is always positive.

2. Properties of Logarithms:

   (A) For $x > 0, y > 0, a > 0$ and $a \neq 1$ the following theorems hold good.

   (i) $\log_a (xy) = \log_a x + \log_a y$

   (ii) $\log_a \left(\dfrac{x}{y} \right) = \log_a x - \log_a y$

   (iii) $\log_a (x^n) = n \log_a x$

   (iv) $\log_{a^n} x = \dfrac{1}{n} \log_a x$

   (v) $\log_{a^n} x^m = \dfrac{m}{n} \log_a x$

   (vi) $\log_a a = 1$

   (vii) $\log_a 1 = 0$

   (B) Base Change Rule

   (i) $\log_y x = \log_z x \log_y z$

   (ii) $\log_y x = \dfrac{1}{\log_x y}$

   (iii) $\log_a x = \dfrac{\log_b x}{\log_b a} = \dfrac{\log x}{\log a}$

   (C) Fundamental logarithmic Identify

   (i) $a^{\log_a x} = x$

   (ii) $a^{\log_b x} = x^{\log_b a}$

   (D) Inequalities (i)

   (i) $\log_a x > 0 \Leftrightarrow (x > 1, a > 1)$ or $(0 < x < 1, 0 < a < 1)$

(ii) $\log_a x < 0 \Leftrightarrow (x > 1, 0 < a < 1)$ *or* $(0 < x < 1, a > 1)$

(E) Inequalities (ii)

(i) If $a > 1$ then $x > y \Leftrightarrow \log_a x > \log_a y$

(ii) If $0 < a < 1$ then $x > y \Leftrightarrow \log_a x < \log_a y$

(iii) If $a > 1$ then $x > a \Leftrightarrow \log_a x > 1$

(iv) If $a > 1$ then $x < a \Leftrightarrow \log_a x < 1$

(v) If $0 < a < 1$ then $x < a \Leftrightarrow \log_a x > 1$

(vi) If $0 < a < 1$ then $x > a \Leftrightarrow 0 < \log_a x < 1$

(vii) If $a > 0, x > 1$ then $\log_a x > 0$

(viii) If $0 < a < 1, x > 1$ then $\log_a x < 0$

(ix) If $0 < a < 1, 0 < x < 1$ then $\log_a x > 0$

(x) If $a > 1, 0 < x < 1$ *then* $> \log_a x < 0$

(xi) If $a > 1$ and $\log_a x > m$ then $x > a^m$.

(xii) If $a > 1$ and $\log_a x < m$ then $x < a^m$

(xiii) If $0 < a < 1$ and $\log_a x < m$, then $x > a^m$

(xiv) If $0 < a < 1$ and $\log_a x > m$ then $x < a^m$

Note: (i) $\log_{10} x$ is usually called a Common Logarithm and is denoted as $\log x$.

(ii) $\log_e x$ is called as Natural logarithm and it is also denoted as $\ln x$.

EXERCISE - 1

1. Solve $\log_{(2x-1)}\left\{\dfrac{x^4 + 2}{2x+1}\right\} = 1$

2. Calculate $81^{1/\log 63} + 27^{\log 9\, 36} + 3^{4/\log 79}$.

3. Find $\log_{30} 8$ if it is known that $\log 5 = a$ and $\log 3 = b$

4   Prove that $\log \dfrac{a+b}{3} = \dfrac{1}{2}(\log a + \log b)$,

    if $a^2 + b^2 = 7\,ab$, $a > 0$, $b > 0$.

5. Solve the equation $x^{1-\log x} = 0.0$.

6. Prove that $\log_p \log_p \dfrac{\sqrt[p]{\sqrt[p]{\ldots\ldots\sqrt[p]{p}}}}{n}$ for $p > 1$.

7. Simplify

    (a) $49^{\frac{1}{\log_5 7}} + 3^{\frac{3}{\log_{\sqrt 6} 3}}$

    (b) $5^{\log_{\frac{1}{5}}\left(\frac{1}{2}\right)} + \log_{\sqrt 2}\dfrac{2}{(\sqrt 7 + \sqrt 3)} + \log_{\sqrt 2}\dfrac{2\sqrt 2}{(\sqrt 7 - \sqrt 3)}$

8. Solve $\sqrt{\log_2 x^4} + 4\log_4 \sqrt{\dfrac{2}{x}} = 2$

9. If $\dfrac{\log a}{c-a} = \dfrac{\log b}{a-b} = \dfrac{\log c}{b-c}$, then prove that $a^b\, b^c\, c^a = 1$.

10. Evaluate: (a) $\log_{10}\tan\left[\left(\dfrac{1024}{2}\right)\pi\right]\log_{10}\tan\left[\left(\dfrac{1024}{2^2}\right)\pi\right]\log_{10}\tan\left[\left(\dfrac{1024}{2^3}\right)\pi\right]\ldots\ldots\log_{10}\tan\left[\left(\dfrac{1024}{2^4}\right)\pi\right]\ldots\ldots\log\tan\left[\left(\dfrac{1024}{2^{20}}\right)\pi\right]$

    (b) $\log_{10}\tan 1° + \log_{10}\tan 2° + \ldots\ldots\ldots\ldots + \log_{10}\tan 89°$

11. Prove that $(0.25)^{\log_{0.5}\left(\frac{1}{4}+\frac{1}{4^2}+\ldots\ldots\infty\right)} = \dfrac{1}{9}$

12. If a, b, c are in G.P. then show that

    $$\dfrac{1}{\log_b x - \log_a x} + \dfrac{1}{\log_b x - \log_c x} = \dfrac{1}{\log_a x} + \dfrac{1}{\log_c x}$$

13. Solve the equation

155

$$\log_{0.5x} x^2 - 14 \log_{16x} x^3 + 40 \log_{4x} \sqrt{x} = 0$$

14. Find all real nimbers x which satisfy the equation

$$2 \log_2 \log_2 x \ \log_{1/2} \log_2 (2\sqrt{2}x) = 1$$

15. Solve the equation $\quad 2 - x + 3 \log_5 2 = \log_5 (3^x - 5^{2-x})$

Simplify the expression $\dfrac{\log_a \sqrt{a^2 - 1} \ \log_{1/a}^2 \sqrt{a^2 - 1}}{\log_{a^2} (a^2 - 1) \log_{\sqrt[3]{a}} \sqrt[6]{a^2 - 1}}$

16. Solve the equation $(\log_2 x)^2 - 5 (\log_2 x) + 6 = 0$

17. Solve the equation
$$\log_3(x^2 - 3x - 5) = \log_3 (7 - 2x).$$

18. Solve the equation
$$\log (x + 4) + \log (2x + 3) = \log (1 - 2x)$$

19. Solve the equation

$$\log_2 (x^2 - 1) = \log_{\frac{1}{2}} (x - 1) \qquad\qquad \text{(i)}$$

20. Solve the equation
$$\log_{x+4} (x^2 - 1) = \log_{x+4} (5 - x). \qquad\qquad \text{(i)}$$

21. Solve the equation $\log^2 x + \log x + 1 = \dfrac{7}{\log \dfrac{x}{10}}$

22. Solve the equation $\log_x (3x^{\log_5 x} + 4) = 2 \log_5 x$

23. Find the least integer 'n' such that $7^n > 10^5$ given that $\log_{10} 343 = 2.5353$.

24. Show that $7 \log \left(\dfrac{16}{15}\right) + 5 \log \left(\dfrac{25}{24}\right) + 3 \log \left(\dfrac{81}{80}\right) = \log 2$.

25. If $\log 3 = 0.4771, \ \log 7 = 0.8450$.

Solve $3^{1+x} = 7^{\frac{x}{2}}$

26. Solve $x + \log(1 + 2^x) = x \log 5 + \log 6$.

27. Solve the equations simultaneously

$$\log_9 x - \log_3 y = 0 ; \qquad x^2 - 14y^2 = 32$$

156

Exercise - 1
Solutions

1. Conditions,

    i) $2x - 1 > 0$, $x \neq 1$

    ii) $2 x' + 1 > 0$,   since $x^4 + 2 > 0$ always.

    $x > \dfrac{1}{2}$ & $x = -1/2$

    $\Rightarrow x > \dfrac{1}{2}$ is domain of equation

    The equivalent form of given equation

    $$\dfrac{x^4 + 2}{2x + 1} = 2x - 1$$

    $$x^4 + 2 = 4x^2 - 1$$

    $$x^4 - 4x^2 + 3 = 0$$

    $$(x^2 - 1)(x^2 - 3) = 0$$

    $$x = \pm 1, \qquad\qquad x = \pm\sqrt{3}$$

    Since $x > 1/2$ and $x \neq 1$

    The only solution is $x = \sqrt{3}$

2. According to (5) we have $81^{1/\log_5 3} = 81^{\log_3 5} = (3^4)^{\log_3 5} = (3^{\log_3 5})^4$

    Using the properties of powers, we obtain $81^{"}$

    According to the definition of a logarithm, we have $3^{\log_3 5} = 5$

    Thus we have $81^{1/\log_5 3} = 5^4 = 625$

    Similarly, $3^{4/\log_7 9} = 3^{4\log_9 7} = (3^2)^{2\log_9 7} = (9^{\log_9 7})^2 = 7^2 = 49$,

    $$27^{\log_9 36} = 27^{\log_3 6} = 3^{3\log_3 6} = (3^{\log_3 6})^3 = 216.$$

    Adding the resulting values together, we get the required number.

    = 890.

    Note : When calculating the values of one logarithmic or exponential expression from some other, known, logarithmic or exponential expressions, we usually factor all the numbers appearing in the given expressions into primes.

3.   We represent $\log_{30} 8$ as $\log_{30} 8 = \dfrac{\log 8}{\log 30}$

Using prime factorization of the numbers 30 and 8 and the properties of logarithms, we obtain

$$\log_{30} 8 = \dfrac{3 \log 2}{\log 5 + \log 3 + \log 2}$$

Taking into account that $\log 2 = \log \dfrac{10}{5} = 1 - \log 5.$

We get $\log_{30} 8 = \dfrac{3(1-a)}{b+1}$

Note :   To prove the identity of two logarithmic expressions when certain conditions are fulfilled, it is sometimes convenient to transform the given conditions and then to take their logarithms.

4.   Let us transform the conditions by isolating the perfect square :
$$a^2 + b^2 + 2ab = 9\, ab,$$
i.e., $(a + b)^2 = 9\, ab$

Taking logarithms of this equality to the base 10 and collecting terms, we obtain

$$2 \log (a + b) - 2 \log 3 = \log a + \log b.$$

$$\Rightarrow \log \dfrac{a+b}{3} = \dfrac{1}{2} (\log a + \log b)$$

then the numbers $\log_a N$, $\log_b N$, $\log_c N$ are three successive terms of an arithmetic progression for any positive value of N.

Note :   When proving identities (i.e., verifying the validity of some equalities on the whole domain of definition of the functions appearing in them), use shoud be made of the same techniques as those employed in simplifying logarithmic and exponential expressions.

5.   The domain of definition of the equation is $x > 0$. In this domain the expression contained on both sides of Equation (18) take on only positive values therefore taking the decimal logarithms of both sides of the equation, we get the equation $\log x^{1-\log x} = \log 0.01$

which is equivalent to equation. (i) $\Rightarrow$ $(1 - \log x) \log x = -2$.

Setting $u = \log x$, we get the equation $(1 - u) u = -2$, $\Rightarrow u_1 = -1$, $u_2 = 2$. It remains to solve the following collection of equations : $\log x = -1$; $\log x = 2$.

From this collection we get : $x_1 = 0.1$, $x_2 = 100$. These are roots of equation (i)

6.   Transforming the irrational expression under the second sign of logarithm, we obtain

$$\frac{\sqrt[p]{\sqrt[p]{\ldots \sqrt[p]{p}}}}{n} = p^{1/pn}$$

Taking logarithms in this equation to the base p, we get

$$\log_p p^{1/pr} = \frac{1}{p^n}$$

Taking logarithms in this expression to the base p once again, we get the required identity.

7.   (a) $y = 49^{\frac{1}{\log_5 7}} + 3^{\frac{3}{\log \sqrt[3]{6}}}$

$$= 49^{\log_7 5} + 3^{3\log_3 \sqrt{6}} = \left(7^{\log_7 5}\right)^2 + \left(3^{\log_3 \sqrt{6}}\right)^3$$

$$= 5^2 + (\sqrt{6})^3 \quad = 25 + 6\sqrt{6}$$

(b) $5^{\frac{\log_{\frac{1}{5}}\left(\frac{1}{2}\right)}{}} + \log_{\sqrt{2}} \frac{2}{(\sqrt{7} + \sqrt{3})} + \log_{\sqrt{2}} \left(\frac{2\sqrt{2}}{(\sqrt{7} - \sqrt{3})}\right)$

$$= 5^{\log_{1/5}\frac{1}{2}} + \log_{\sqrt{2}} \frac{4\sqrt{2}}{4}$$

$$= 5^{\log_{1/5}\frac{1}{2}} + 1 = 5^{\log_5 2} + 1 = 2 + 1 = 3$$

Note that, $\log 1/5 \ \frac{1}{2} = \frac{\log_5 \frac{1}{2}}{\log_5 \frac{1}{5}} = \frac{-\log_5 2}{-\log_5 5} = \log_5 2.$

Problems in Algebra

8. $\sqrt{\log_2 x^4} + 4\log_4 \sqrt{\dfrac{2}{x}} = 2$

$2\sqrt{\log_2 x} + 4\log_{2^2} \sqrt{\dfrac{2}{x}} = 2$

$2\sqrt{\log_2 x} + \dfrac{4}{2}\cdot\dfrac{1}{2}\log_2\left(\dfrac{2}{x}\right) = 2$

$2\sqrt{\log_2 x} + 1 - \log_2 x = 2$

Let $t = \log_2 x$

$2\sqrt{t} + 1 - t = 2$

$2\sqrt{t} + t + 1$

Squaring we get, $4t = (t+1)^2$

$\Rightarrow (t-1)^2 = 0$

$\Rightarrow t = 1$

$\Rightarrow \log_2 x = 1$

$\Rightarrow x = 2$

Conditions applied :

i) $x > 0$ (for $\log_2 x^4$ to be defined)

ii) $\dfrac{2}{x} \geq 0 \Rightarrow x > 0$ (for $\sqrt{\dfrac{2}{x}}$ to be defined.)

iii) $\log_2 x^4 \geq 0 \Rightarrow x^4 \geq 1$ which is true for $x \geq 1$ or $\leq x -1$.

The final condition to be satisfied is $x > 1$

$\therefore$ Domain of equation is $(0, \infty)$ and the solution $x = z \in (0, \infty)$

(In logarithmic equation find the solution and then check whether is satisfies all the conditions i.e. whether it belongs to the domain of the equation).

9. Let, $\dfrac{b\log a}{b(c-a)} = \dfrac{c\log b}{c(a-b)} = \dfrac{a\log c}{a(b-c)} = k$

$\log a^b = k\,(bc - ba)$

160

$\log b^c = k \ (ca - cb)$

$\log c^a = k \ (ab - ac)$

Adding, $\log (a^b \ b^c \ c^a) = k \ (0) \Rightarrow a^b \ b^c \ c^a = 1$

10. (a) An arbitary term of $\log \left[\tan \left(\dfrac{1024}{2^n} \right)^\pi \right]$

Note that, $1024 = 2^{10}$

For $n = 12$, we have, $\log \left[\tan \left(\dfrac{2^{10}}{2^{12}} \right) \pi \right] = \log \left(\tan \dfrac{\pi}{4} \right) = \log 1 = 0$

∴ The expression above is product and one of the term is zero.

∴ Hence the solution is zero.

(b) $\log_{10} \tan 1^0 + \log_{10} \tan 2^0 + \ldots\ldots + \log_{10} \tan 89^0$

Note that : $\tan 89^0 = \cot 1^0$

Similarly for all.

∴ $\log_{10} (\tan 1^0) + \log(\tan 2^0) + \log \tan(45^0) + \ldots\ldots + \log (\cot 2^0) + \log(\cot 1^0)$

$= \log_{10} \left[(\tan 1°)(\tan 2°)\ldots(\tan 45°). \dfrac{1}{\tan(44°)} \cdot \dfrac{1}{\tan(43°)} \ldots \dfrac{1}{\tan 1°} \right]$

$= \log_{10}[1] = 0$

11. L.H.S. $= (0.25)^{\log_{0.5} \left(\frac{1}{4} + \frac{1}{4^2} + \ldots \infty \right)}$

Let $S = \dfrac{1}{4} + \dfrac{1}{4^2} + \ldots \infty = \dfrac{1/4}{1 - \dfrac{1}{4}} = \dfrac{1}{3}$

[Summation of an infinite G.P. $a, ar, ar^2, \ldots\ldots \infty$ is $S = \dfrac{a}{1-r}$]

L.H.S. $= (0.25)^{\log 0.5 \frac{1}{3}} = \left[\left(\dfrac{5}{10} \right)^2 \right]^{\log 1/2 \frac{1}{3}}$

161

$$= \left[\left(\frac{5}{10} \right)^2 \right]^{\log 5/10 \frac{1}{3}}$$

$$= \left(\frac{5}{10} \right)^{\log 5/10 (1/3)^2} = \frac{1}{9} = \text{R.H.S.}$$

12.   a, b, c are in G.P.

$$\therefore \qquad ac = b^2.$$

$$\text{L.H.S.} \quad = \frac{1}{\log_b x - \log_a x} + \frac{1}{\log_b x - \log_c x}$$

$$= \frac{1}{\dfrac{1}{\log_x b} - \dfrac{1}{\log_x a}} + \frac{1}{\dfrac{1}{\log_x b} - \dfrac{1}{\log_x c}}$$

$$= \frac{\log_x b . \log_x a}{\log_x a - \log_x b} + \frac{\log_x b . \log_x c}{\log_x c - \log_x b}$$

$$= \frac{\log_x b \log_x a}{-\log_x r} + \frac{\log_x b . \log_x c}{\log_x r}$$

$$= \frac{1}{\log_x r} \left[\log_x b(\log_x c / a) \right] = 2 \log_x b$$

$$\text{R.H.S.} \quad = \frac{1}{\log_a x} + \frac{1}{\log_c x}$$

$$= \log_x a + \log_x c$$

$$= \log_x a.c = 2 \log_x b = \text{R.H.S.}$$

13.   Changing the base of all the logarithms to 2, we get :

$$\frac{\log_2 x^2}{\log_2 0.5x} - \frac{14 \log_2 x^3}{\log_2 16x} + \frac{40 \log_2 \sqrt{x}}{\log_2 4x} = 0 \text{ and further}$$

$$\Rightarrow \frac{2\log_2 |x|}{\log_2 x - 1} - \frac{42\log_2 x}{\log_2 x + 4} + \frac{20\log_2 x}{\log_2 x + 2} = 0$$

From the given equation it follows that $x > 0$, and therefore $|x| = x$, that is, $\log_2 |x| = \log_2 x$. Setting $u = \log_2 x$, we get the equation

$$\frac{2u}{u-1} - \frac{42u}{u+4} + \frac{20u}{u+2} = 0$$

whose roots are : $u_1 = -\dfrac{1}{2}$, $u_2 = 0$, $u_3 = 2$.

Now, the problem is reduced to solving the following collection of equations :

$$\log_2 x = -\frac{1}{2}; \ \log_2 x = 0; \ \log_2 x = 2.$$

From the first equation we get : $x_1 = \dfrac{\sqrt{2}}{2}$, from the second : $x_2 = 1$,

from the third : $x_3 = 4$

All the found values are the roots of the original equation (the reader is invited to make this sure independently)

14. $\log_2(\log_2 x)^2 - \log_2 (\log_2 (2^{3/2} x) = 1$

$\Rightarrow \log_2(\log_2 x)^2 - \log_2 [\log_2 (2^{3/2}) + \log_2 x]$

$$\log_2\left[\frac{\left(\log_2 x\right)^2}{\frac{3}{2} + \log_2 x}\right] = 1$$

$$\frac{\left(\log_2 x\right)^2}{\frac{3}{2} + \log_2 x} = 2$$

Let $= \log_2 x$

$$\frac{t^2}{\frac{3}{2} + t} = 2 \qquad\qquad t \neq -3/2$$

Problems in Algebra

$$t^2 = 2\left(\frac{3}{2} + t\right)$$

$$t^2 = 3 + 2t$$

$$t = \frac{+2 \pm \sqrt{4+12}}{2}, \qquad t = \frac{2 \pm 4}{2} = 3 \text{ or } -1$$

$$x = 2^3 = 8 \qquad \text{or} \qquad x = 2^{-1} = 1/2$$

Moreover conditions, for the equation to be defined, $\log_2 x > 0 \Rightarrow$ $x > 1$

$\therefore$ we need to discard $x = 1/2$

The only solution is $x = 8$

$$\log_{\frac{1}{\pi}}\left(x^2 + x - 2\right) > \log_{\frac{1}{\pi}}\left(x + 3\right)$$

$$\therefore \pi > 1 \Rightarrow \frac{1}{\pi} < 1$$

Conditions : i) $x + x - 2 > 0$

$\quad (x+2)(x-1) > 0$

$\quad x > 1 \text{ x } x < -2$

ii) $x + 3 > 0 \Rightarrow x > -3$

Now the equivalent equation is $x^2 + x - 2 < x + 3$

$$x^2 < 5 \quad -\sqrt{5} < x < \sqrt{5}$$

The solution is $x \in \left(-\sqrt{5}, -2\right) \cup \left(1\sqrt{5}\right)$

15. According to (3), we have

$$(\log_{1/a}\sqrt{a^2-1})^2 = (\log_a\sqrt{a^2-1})^2 = (\log_a \sqrt{a^2-1})^2, \qquad \dots (*)$$

$$\log_{\sqrt[3]{a}}\sqrt[6]{a^2-1} = \log_{(\sqrt[3]{a})3}(\sqrt[6]{a^2-1})^3 = \log_a\sqrt{a^2-1}, \qquad \dots (**)$$

$$\log_{a2}(a^2-1) = \log_{(a^2)1/2}(a^2-1)^{1/2}\log_a\sqrt{a^2-1}. \qquad \dots (***)$$

Substituting the right-hand sides of expressions (*) to (***) into the initial fraction, we obtain

$$\frac{\log_a \sqrt{a^2-1}\ \log_a^2 \sqrt{a^2-1}}{\log_a \sqrt{a^2-1}\ \log_a \sqrt{a^2-1}} = \log_a \sqrt{a_2-1}$$

16. Designating $\log_2 x = y$, we get an equation
$$y^2 - 5y + 6 = 0$$
whose roots are $y_1 = 2, y_2 = 3$. Thus the initial equation is equivalent to two equations.
$$\log_2 x = 2, \log_2 x = 3,$$
whose solutions are $x = 4$ and $x = 8$.

17. By Theorem 1, Equation (8) is equivalent to the following mixed system :
$$\begin{cases} x^2 - 3x - 5 = 7 - 2x \\ x^2 - 3x - 5 > 0 \\ 7 - 2x > 0 \end{cases}$$

Solving the equation of this system, we get: $x_1 = 4, x_2 = -3$. Of these two values only $x = -3$ satisfies both inequalities of System (9) (that is, the value $x = 4$ does not belong to the domain of definition of Equation (i)). Therefore, $x = -3$ is the solution of Equation (i).

18. We transform Equation (i) to the form
$$\log((x+4)(2x+3)) = \log(1-2x), \text{ and further}$$
$(x+4)(2x+3) = 1 - 2x.$
(ii) From Equation (ii) we find : $x_1 = -1, x_2 = -5.5$.
The domain of definition of Equation (ii) is given by the system of inequalities :
$$\begin{cases} x + 4 > 0 \\ 2x + 3 > 0 \\ 1 - 2x > 0 \end{cases} \quad \text{(iii)}$$

Substituting the found roots of Equation (11) into System (12), we make sure that $x_1 = -1$ satisfies this system, while $x_2 = -5.5$ does not. Thus, $x = -1$ is the only root of Equation (i).

19. First of all, let us pass in Equation (i) to logarithms with equal bases. Since $\log_a N = \log_{a^k} N^k$, Equation (i) is transformed to the

Problems in Algebra

following :

$$\log_2(x^2 - 1) = \log_{(2)^{-1}}(x - 1)^1$$

$$\Rightarrow \log_2(x^2 - 1) = \log_2(x - 1)$$

$$\Rightarrow \log_2(x^2 - 1) = \log_2 \frac{1}{x - 1} \qquad\qquad \text{(ii)}$$

Solving Equation (ii), we find:

$$x_1 = 0, \ x_2 = \frac{1 + \sqrt{5}}{2}, \ x_3 = \frac{1 - \sqrt{5}}{2}$$

It remains only to choose from the found values those which satisfy

the system of inequalities $\begin{cases} x^2 - 1 > 0 \\ x - 1 > 0 \end{cases}$.

Solving this system, we find that $x > 1$. Of the found values x_1, x_2,

x_3, only $x_2 = \dfrac{1 + \sqrt{5}}{2}$ satisfies the inequality $x > 1$. Hence, $x = \dfrac{1 + \sqrt{5}}{2}$

is the only root of Equation (i).

20.　By Theorem 2, this equation is equivalent to the system

$$\begin{cases} x^2 - 1 = 5 - x \\ x^2 - 1 > 0 \\ 5 - x > 0 \\ x + 4 > 0 \\ x + 4 \neq 1 \end{cases} \qquad \text{(ii)}$$

Solving the equation, entering System (ii), we get : $x_1 = 2$, $x_2 = -3$. Of these two values only $x = 2$ satisfies the rest of the conditions of System (ii). Thus, $x = 2$ is a root of Equation (i).

21.　Since $\log \dfrac{x}{10} = \log x - 1$, the given equation can be rewritten :

$$\log^2 x + \log x + 1 = \frac{7}{\log x - 1} .$$

166

Setting $u = \log x$, we get the equation $u^2 + u + 1 = \dfrac{7}{u-1}$,

whence we find : $u = 2$. From the equation $\log x = 2$ we find : $x = 100$. This is just the only root of the original equation.

22. Using the definition of logarithm, we transform equation

$$x^{2\log_5 x} = 3x^{\log_5 x} + 4$$

Setting $u = x^{\log_5 x}$, we get the equation $u^2 - 3u - 4 = 0$, whose roots are : $u_1 = -1$, $u_2 = 4$. Now, the problem is reduced to solving the following collection of equations : $x^{\log_5 x} = -1$; $x^{\log_5 x} = 4$. Since $x^{\log_5 x} > 0$, and $-1 < 0$, the first equation of this collection has no solution. Taking the logarithms to the base 5 of both sides of the second equation we get :

$$\log_5^2 x = \log_5 4, \text{ i.e. } \log_5 x = \pm \sqrt{\log_5 4}\,.$$

$\Rightarrow x_{1,2} = 5 \pm \sqrt{\log_5^4}$. These are roots of equation.

23. $7^n > 10^5$.

Taking log on both sides $n\log_{10} 7 > 5\log_{10} 10$

$\therefore \qquad n\log_{10} 7 > 5 \qquad\qquad\qquad ... \text{(i)}$

Given $\log_{10} 343 = 2.5353$

$\log_{10} 7^3 = 2.5353$

$3\log_{10} 7 = 2.5353 \quad \Rightarrow \quad \log_{10} 7 = 0.8451$

from eq. (i) $\qquad \Rightarrow \qquad n(0.8451) > 5$

$\Rightarrow \qquad n > \dfrac{5}{0.8451} = 5.916$

$\Rightarrow \qquad n = 6$.

24. $LHS = \log\left(\dfrac{16}{15}\right)^7 + \log\left(\dfrac{25}{24}\right)^5 + \log\left(\dfrac{81}{80}\right)^3$

$$= \log \left[\left(\frac{16}{15} \right)^7 \left(\frac{25}{24} \right)^5 \left(\frac{81}{80} \right)^3 \right]$$

$$= \log \left[\left(\frac{2^4}{3 \times 5} \right)^7 \cdot \left(\frac{5^2}{2^3 \times 3} \right)^5 \left(\frac{3^4}{2^4 \times 5} \right)^3 \right]$$

$$= \log \left[\frac{2^{28}}{3^7.5^7} \cdot \frac{5^{10}}{2^{15}.3^5} \cdot \frac{3^{12}}{2^{12}.5^3} \right]$$

$$= \log \left[\frac{2^{28}}{2^{27}} \right] = \log 2 .$$

25. $3^{1+x} = 7^{\frac{x}{2}}$

Taking log on both sides

$$\log 3^{1+x} = \log 7^{\frac{x}{2}}$$

$$(1+x)\log 3 = \frac{x}{2}\log 7$$

$$x\left(\log 3 - \frac{\log 7}{2} \right) = -\log 3$$

$$\Rightarrow \quad x\left(\frac{2\log 3 - \log 7}{2} \right) = -\log 3$$

$$\Rightarrow \quad x = \frac{2\log 3}{\log 7 - 2\log 3} = \frac{2(0.4771)}{0.8450 - 2(0.4771)}$$

$$\therefore \quad x \simeq -8.7$$

26.   The equation is

$$x + \log_{10}(1+2^x) = x\log_{10} 5 + \log_{10} 6$$

$$x = \log_{10} 5^x + \log_{10} 6 - \log_{10}(1+2^x)$$

$$x = \log_{10} \left(\frac{5^x.6}{1+2^x} \right)$$

$$\Rightarrow \quad 10^x = \frac{5^x \cdot 6}{1+2^x}$$

$$\Rightarrow \quad (5 \times 2)^x = \frac{5^x \cdot 6}{1+2^x}$$

$$5^x \cdot 2^x = \frac{5^x \cdot 6}{1+2^x}$$

5^x cannot be zero for any finite x.

$$\Rightarrow \quad 2^x = \frac{6}{1+2^x}$$

$$\Rightarrow \quad 2^x(1+2^x) = 6$$

$$(2^x)^2 + 2^x - 6 = 0$$

$$(2^x + 3)(2^x - 2) = 0$$

$$\Rightarrow \quad 2^x = -3 \qquad or \qquad 2^x = 2$$

2^x cannot be negative $\Rightarrow 2^x = 2$

$$\Rightarrow \quad x = 1.$$

27. $$\log_9 x - \log_3 y = 0$$

$$\log_9 x = \log_3 y$$

$$\frac{\log x}{\log 9} = \frac{\log y}{\log 3}$$

$$\frac{\log x}{\log 3^2} = \frac{\log y}{\log 3}$$

$$\frac{\log x}{2 \log 3} = \frac{\log y}{\log 3}$$

$$\log x = 2 \log y$$

$$\Rightarrow \quad \log x = \log y^2$$

$$\Rightarrow \quad x = y^2$$

Second equation is

169

$$x^2 - 14y^2 = 32$$

$$y^4 - 14y^2 - 32 = 0$$

$$(y^2 - 16)(y^2 + 2) = 0$$

$\Rightarrow \qquad y^2 = 16 \qquad or \qquad y^2 = -2$

$\Rightarrow \qquad y = \pm 4$

y cannot be -4

$\because \qquad \log_3 y$ is not defined.

$\therefore \qquad y = 4$

$\therefore \qquad$ when $y = 4$, $x = 16$

EXERCISE - 2   [Time: 3hrs]

1. If $\dfrac{\log a}{b-c} = \dfrac{\log b}{c-a} = \dfrac{\log c}{a-b}$. Prove that $a^a . b^b . c^c = 1$

2. Find x if $\log_{\sqrt{8}} x = \dfrac{10}{3}$.

3. Find x if $\log_e 2 \log_x 625 = \log_{10} 16 \log_e 10$

4. If n is a natural number such that $n = P_1^{\alpha_1} P_2^{\alpha_2} \dots P_k^{\alpha_k}$ where $P_1, P_2, \dots P_k$ are distinct primes, show that $\log n \geq K \log 2$.

5. If $\log_{0.3}(x-1) < \log_{0.09}(x-1)$. Find the interval in which x lies.

6. Find the least value of the expression $(2 \log_{10} x - \log_x 0.01)$ for $x > 1$.

7. If $x = 1 + \log_a bc$; $\quad y = 1 + \log_b ca$; $\quad z = 1 + \log_c ab$; then prove that $xyz = xy + yz + zx$

8. If $\log\left(\dfrac{a+b}{3}\right) = \dfrac{1}{2}(\log a + \log b)$. Prove that $\dfrac{a}{b} + \dfrac{b}{a} = 7$.

9. If $a^x = b^y = c^z = d^w$. Prove that $\log_a (bcd) = x\left(\dfrac{1}{y} + \dfrac{1}{z} + \dfrac{1}{w}\right)$.

Exercise - 2

Answers

2.   32

3.   5

5. $(2, \infty)$

6.   4

BLANK PAGE LEFT INTENTIONALLY

9. PROGRESSIONS

1. Arithmetical Progression

   Definition: Let $'a', 'd'$ be two real numbers. Then a $a, a+d, a+2d, a+3d, ...$ are said to be in **arithmetical progression** (A.P.).

   Where $'a'$ is called the first term of the progression, and d is common difference.

   Let $'T_n'$ denote the nth term of the progression, then

   $$T_n = a + (n-1)d$$

   If there are $'n'$ terms of an AP of which $'a'$ is the first term and $'l'$ is the last term then the sum of the series given by

   $$S_n = \frac{n}{2}[a+l]$$

   If there are $'n'$ terms of an AP of which $'a'$ is the first term and $'d'$ is the common difference then the sum of the series is given by

   $$S_n = \frac{n}{2}[2a + (n-1)d]$$

2. Geometric Progression

   Definition: Let $'a', 'r'$ be two real numbers then $a, ar, ar^2, ...$ are said to be in Geometric Progressions (GP).

   Here $'a'$ is called the first term, and r is the common ratio.

   Let $'T_n'$ denote the nth term of the progression, then

   $$T_n = ar^{n-1}$$

   If S_n denotes the sum to $'n'$ terms

   $$S_n = a + ar + ... + ar^{n-1}$$

   then S_n is given by,

   $$S_n = \frac{a(1-r^n)}{1-r}$$

   Note: If $|r| < 1$ and when the number of terms of the progression tends to infinity then $|r|^n \rightarrow 0$

$$\therefore \qquad S_\infty = \frac{a}{1-r}$$

3. **Harmonic Progression**

When a set of reals are in Arithmetical Progression then their reciprocals are said to be in Harmonic Progression (HP)

Thus a HP is of the form

$$\frac{1}{a}, \frac{1}{a+d}, \frac{1}{a+2d}, \dots\dots$$

If T_n denotes the nth term of the HP

then $\qquad T_n = \dfrac{1}{a+(n-1)d}$

4. **Arithmetic Means**

Let a, b be two reals. A is called the arithmetic mean between a, b. If a, A, b are in Arithmetical Progression, then

$$A = \frac{a+b}{2}$$

Let A_1, A_2 be two arithmetic means between a, b.

Then a, A_1, A_2, b are in A.P.

$\dfrac{2a+b}{3}$ *and* $\dfrac{a+2b}{3}$ are the two arithmetic means between a, b.

5. **Geometric Means**

Let a, b be two positive reals. G is called the geometric mean between a, b if a, G, b are in geometric progression.

Then, $\qquad G = \sqrt{ab}$

Let a, G_1, G_2, b are in G.P.

$a^{\frac{2}{3}} b^{\frac{1}{3}}$ and $a^{\frac{1}{3}} b^{\frac{2}{3}}$ are the two GMs between the two positive quantities a, b.

6. **Harmonic Means**

Let be two reals. H is called the harmonic mean (HM) between a, b

is defined as a, H, b are in HP

Then, $$H = \frac{2ab}{a+b}$$

7. Relation between A, G, H

Let a, b be two positive reals.

Then we have $A = \dfrac{a+b}{2}$

$G = \sqrt{ab}$

$H = \dfrac{2ab}{a+b}$

Then $\quad A > G > H$

or $\quad H < G < A$

EXERCISE - 1

1. Find the sum of integers from 1 to 100 that are divisible by 2 or 5.

2. The sums of n terms of two APs are in the ratio $7n+1 : 4n+27$. Find the ratio of their n^{th} terms.

3. If $\alpha_1, \alpha_2, \ldots\ldots, \alpha_n$ are in AP where $\alpha_i > 0$ for each i, show that

$$\frac{1}{\sqrt{\alpha_1}+\sqrt{\alpha_2}} + \frac{1}{\sqrt{\alpha_2}+\sqrt{\alpha_3}} + \ldots\ldots + \frac{1}{\sqrt{\alpha_{n-1}}+\sqrt{\alpha_n}} = \frac{n-1}{\sqrt{\alpha_n}+\sqrt{\alpha_1}}$$

4. If the harmonic mean of two numbers is that to their geometric mean is 12 : 13, Prove that the numbers are in the ratio 4 : 9 or 9 : 4.

5. Find the greatest value of positive integers ' n ' such that the sum upto ' n ' terms of the series $1 + \dfrac{1}{2} + \dfrac{1}{2^2} + \ldots\ldots < 2 - \dfrac{1}{1000}$

6. If 'n' is positive integer then, prove that

$\underbrace{111\ldots\ldots 1}_{2n \ times} - \underbrace{222\ldots\ldots 2}_{n \ times}$ is the square of positive integer

7. A square is drawn by joining the midpoints of the sides of a given square. A third square is drawn inside the second square in the same

175

way and this process continues indefinitely. If the side of the first square is 4 units determine the sum of the areas of the squares.

8. Find the value of m so that $\dfrac{a^{m+1} + b^{m+1}}{a^m + b^m}$ is the G.M. between a and b

9. If the pth, q^{th} and r^{th} term of an A.P. be a, b, c respectively; show that $\quad a(q-r) + b(r-p) + c(p-q) = 0$

10. The first term of an arithmetical progression is $\log a$ and the second term is $\log b$. Express the sum to 'n' terms as a logarithm.

11. Determine all positive finite series in Geometric Progression with first term '1', common ratio be an integer greater than '1' and sum is 2002.

12. If x, y and z are positive real numbers different from 1, and $x^{18} = y^{21} = z^{28}$, show that 3, 3 $\log_y x$, 3 $\log_z y$, 7 $\log_x z$ are in A.P.

13. Find the coefficient of x^{98} in the expansion of
    $(x + 1)(x + 2)(x + 3) \dots \dots (x + 100)$

14. mn squares of equal size are arranged to form a rectangle of dimensions m by n, where m and n are natural numbers. Two square will be called 'neighbours' if they have exactly one common side. A natural number is written in each square such that the number written in any square is the arithmetic mean of the numbers written in its neighbouring squares. Show that this is possible only if all the natural numbers are equal.

15. If $a_1, a_2, a_3 \dots \dots a_n$ are in A.P. with common difference d, then sin d[cosec a_1 cosec a_2 + cosec a_2 cosec a_3 + $\dots \dots$ + cosec a_{n-1} cosec a_n] = cot a_1 − cot a_n

16. If $\sqrt[x]{a} = \sqrt[y]{b} = \sqrt[z]{c}$ and if a, b, c are in G.P.,
    then prove that x, y, z are in A.P.

17. If a, b, c are in G.P. then the equation $ax^2 + 2bx + c = 0$ and $dx^2 + 2ex + f = 0$ have a common root if $\dfrac{d}{a}, \dfrac{e}{b}, \dfrac{f}{c}$ are in A.P.

18. If $a^x = b^y = c^z$ and x, y, z are in G.P. show that $\log_b a = \log_c b$.

19. Sum the series $1 + 3x + 5x^2 + 7x^3 + \dots$
    (a) to n terms

(b) to infinity where |x| < 1

20. Let p be the first term of n AM's between two numbers and q be first of n HM's between the same two numbers. Prove that hte value of q can not lie between p and $\left(\dfrac{n+1}{n-1}\right)^2$ P.

21. Sum the series 1 + 4 + 10 + 22 + 46 + ... to n terms

22. Evaluate $\displaystyle\sum_{k=1}^{n}\tan^{-1}\left(\dfrac{2k}{2+k^2+k^4}\right)$

23. Evaluate sum of n terms of the series $\dfrac{8}{5}+\dfrac{16}{65}+\dfrac{24}{325}+$ ....

24. Let x = 1 + 3a + 6a² + 10a³ + ... | a | < 1

    y = 1 + 4b + 10b² + 20b³ .... | b | < 1

    Find s = 1 + 3 (ab) + 5 (ab)² + ...

    In terms of x, y.

25. If a > 0, b > 0, c > 0, prove that

    $$\dfrac{1}{s-a}+\dfrac{1}{s-b}+\dfrac{1}{s-c}>\dfrac{9}{2s}$$

    where s = a + b + c

26. Find a + ar + ar² .... ∞ where a is the value of x for which the function f(x) = 7 + 2x ln 25 – 5^{x–1} – 5^{2-x} has the greatest value and r is the limit of $\displaystyle\lim_{x\to0}\int_{0}^{x}\dfrac{t^2 dt}{\{x^2\tan(T_1+x)\}}$.

27. Consider the A.Ps 17, 21, 25, ... and 16, 21, 26 ..... Find the sum of first 100 common terms appearing in the two series.

28. Sum the series 1.2.3 + 2.3.4 + 3.4.5 + ... n terms

29. The sum of squares of three distinct real numbers which are in GP is s². If their sum is a s, show that $a^2\in\left[\dfrac{1}{3},1\right]\cup(1,3)$.

30. The consecutive digits of a three digit number are in G.P. if the middle digit be increased by 2 then they form an A.P. If 792 is subtracted from this number then we get the number consisting of

same three digits but in reverse order. Find the number.

31.   Find the sum of first 24 terms of an A.P. given by $a_1, a_2, a_3 ..... a_{24}$ if it is known that $a_1 + a_5 + a_{10} + a_{15} + a_{20} + a_{24} = 225$

32.   . . . . $^2 : n^2$. Show that the ratio of the mth and nth terms is $(2m-1) : (2n-1)$

33.   Find the sum of the following series $5 + 55 + 555 + ... $ to n terms.

34.   If a, b, c are in A.P., α, β, γ are in H.P. and $a\alpha, b\beta, c\gamma$ are in G.P.,

Prove that $a : b : c = \dfrac{1}{\gamma} : \dfrac{1}{\beta} : \dfrac{1}{\alpha}$

35.   Find the coefficient of x^{n-1} in the expansion of $(1 + 2x + 3x^2 + ... nx^{n-1})^2$

36.   If $x = 1 + a + a^2 + ... + \infty$ ($|a| < 1$) and $y = 1 + b + b^2 + ... + \infty$ $(|b| < 1)$

       then prove that $1 + ab + a^2 b^2 + ... = \dfrac{xy}{x + y - 1}$

EXERCISE - 1

Solutions

1.   Integers that are divisible by 2 are 2 4, 6, ..., 100.

      Sum = $2 + 4 + 6 + ... + 100 = \dfrac{50}{2}(2 + 100) = 2550$.

      Integers that are divisible by 5

               $5, 10, 15, ..., 100$.

      Sum = $5 + 10 + 15 + ... + 100 = \dfrac{20}{2}(5 + 100) = 1050$.

      Integers that are divisible by both 2 and 5,

               $10, 20, ..., 100$.

      Sum = $10 + 20 + ... + 100 = \dfrac{10}{2}[10 + 100] = 550$.

      ∴   Sum of the integers that are divisible by 2 or 5.

$$= 2550 + 1050 - 550 = 3050 \cdot$$

2. Let a_1, a_2 and d_1, d_2 be the first terms and common differences of the two APs.

Given $= \dfrac{S_n}{S_n^1} = \dfrac{7n+1}{4n+27}$

$$\dfrac{\dfrac{n}{2}[2a_1 + (n-1)d_1]}{\dfrac{n}{2}[2a_2 + (n-1)d_2]} = \dfrac{7n+1}{4n+27}$$

$$\Rightarrow \quad \dfrac{2a_1 + (n-1)d_1}{2a_2 + (n-1)d_2} = \dfrac{7n+1}{4n+27} \qquad \dots (1)$$

$$\dfrac{T_n}{T_n^1} = \dfrac{a_1 + (n-1)d_1}{a_2 + (n-1)d_2} = \dfrac{2[a_1 + (n-1)d_1]}{2[a_2 + (n-1)d_2]} \qquad \dots (2)$$

(1) gives $\dfrac{2a_1 + (n-1)d_1}{2a_2 + (n-1)d_2} = \dfrac{7n-1}{4n+27}$

to get (2) we find that instead of $(n-1)$ in (1).

We have $2(n-1)$ in (2).

Replace $(n-1)$ by $2(n-1)$ in (1)

Again (1) gives

$$\dfrac{2a_1 + (n-1)d_1}{2a_2 + (n-1)d_2} = \dfrac{7(n-1)+8}{4(n-1)+31}$$

$$\therefore \quad \dfrac{2a_1 + 2(n-1)d_1}{2a_2 + 2(n-1)d_2} = \dfrac{7 \times 2(n-1)+8}{4 \times 2(n-1)+31}$$

$$= \dfrac{14(n-1)+8}{8(n-1)+31} = \dfrac{14n-14+8}{8n-8+31}$$

$$= \dfrac{14n-6}{8n+23}$$

3. $\alpha_1, \alpha_2, \dots, \alpha_n$ are in AP

179

Let $'d'$ be the common difference of AP.

Then $\alpha_2 - \alpha_1 = d$, $\alpha_3 - \alpha_2 = d$,......, $\alpha_n - \alpha_{n-1} = d$ ... (1)

and $\alpha_n - \alpha_1 = (n-1)d$ ... (2)

Let us take the first term $\dfrac{1}{\sqrt{\alpha_1} + \sqrt{\alpha_2}}$

$$\dfrac{1}{\sqrt{\alpha_1} + \sqrt{\alpha_2}} = \dfrac{\sqrt{\alpha_1} - \sqrt{\alpha_2}}{\left(\sqrt{\alpha_1} + \sqrt{\alpha_2}\right)\left(\sqrt{\alpha_1} - \sqrt{\alpha_2}\right)}$$

$$= \dfrac{\sqrt{\alpha_1} - \sqrt{\alpha_2}}{\alpha_1 - \alpha_2} = \dfrac{\sqrt{\alpha_1} - \sqrt{\alpha_2}}{-d}$$

L.H.S $= \dfrac{\sqrt{\alpha_1} - \sqrt{\alpha_2}}{d} + \dfrac{\sqrt{\alpha_2} - \sqrt{\alpha_3}}{-d} + ...+ \dfrac{\sqrt{\alpha_{n-1}} - \sqrt{\alpha_n}}{-d}$

Similarly, $\dfrac{1}{\sqrt{\alpha_2} + \sqrt{\alpha_3}} = \dfrac{\left(\sqrt{\alpha_2} - \sqrt{\alpha_3}\right)}{-d}$

$$= -\dfrac{1}{d}\left[\sqrt{\alpha_1} - \sqrt{\alpha_2} + \sqrt{\alpha_2} - \sqrt{\alpha_3} + ...+ \sqrt{\alpha_{n-1}} - \sqrt{\alpha_n}\right]$$

$$= -\dfrac{1}{d}\left[\sqrt{\alpha_1} - \sqrt{\alpha_n}\right]$$

$$= -\dfrac{\left(\sqrt{\alpha_1} - \sqrt{\alpha_n}\right)\left(\sqrt{\alpha_1} + \sqrt{\alpha_n}\right)}{d\left(\sqrt{\alpha_1} + \sqrt{\alpha_n}\right)}$$

$$= \dfrac{-\left(\alpha_1 - \alpha_n\right)}{d\left(\sqrt{\alpha_1} + \sqrt{\alpha_n}\right)} = \dfrac{(n-1)d}{d\left(\sqrt{\alpha_1} + \sqrt{\alpha_n}\right)} \qquad \text{using (2)}$$

$$= \dfrac{n-1}{\sqrt{\alpha_1} + \sqrt{\alpha_n}} = RHS \ .$$

4.   Let a, b be the two numbers.

The harmonic mean $= \dfrac{2ab}{a+b}$

The geometric mean $= \sqrt{ab}$

Given $\dfrac{\left(\dfrac{2ab}{a+b}\right)}{\sqrt{ab}} = \dfrac{12}{13}$

$\dfrac{\sqrt{ab}}{a+b} = \dfrac{6}{13};$ $\qquad a+b = \dfrac{13}{6}\sqrt{ab}$

Dividing by $\sqrt{ab}$ throughout $\sqrt{\dfrac{a}{b}} + \sqrt{\dfrac{b}{a}} = \dfrac{13}{6}$

Let $\qquad \sqrt{\dfrac{a}{b}} = t$

$\qquad t + \dfrac{1}{t} = \dfrac{13}{6}.$

$\Rightarrow \qquad 6t^2 - 13t + 6 = 0$

$\qquad (3t-2)(2t-3) = 0$

$\qquad t = \dfrac{2}{3} \qquad or \qquad \dfrac{3}{2}$

$\qquad \sqrt{\dfrac{a}{b}} = \dfrac{2}{3} \qquad or \qquad \dfrac{3}{2}$

$\Rightarrow \qquad \dfrac{a}{b} = \dfrac{4}{9} \qquad or \qquad \dfrac{9}{4}$

$\therefore \qquad a:b = 4:9 \quad or \quad 9:4.$

5. $\quad S_n = \dfrac{a(1-r^n)}{1-r} = \dfrac{1\left(1-\left(\dfrac{1}{2}\right)^n\right)}{1-\dfrac{1}{2}} < 2 - \dfrac{1}{1000}$

$\Rightarrow \qquad 2\left(1-\dfrac{1}{2^n}\right) < 2 - \dfrac{1}{10^3}$

$$\Rightarrow \quad 2-\frac{1}{2^{n-1}} < 2-\frac{1}{10^3} \Rightarrow \frac{1}{2^{n-1}} > \frac{1}{10^3}$$

$$\Rightarrow \qquad 2^{n-1} < 10^3 = 1000$$

$$\Rightarrow \qquad \frac{2^n}{2} < 1000 \Rightarrow 2^n < 2000$$

for $n=1$ *to* $10 \qquad 2^n < 2000 \qquad\qquad$ ... (1)

for $n=11$, $2n = 2048 > 2000$

$\therefore$ Greatest integer satisfying (1) is $n=10$.

6. $\underbrace{111........1}_{2n \ times} - \underbrace{222........2}_{n \ times}$

$= 1\times10^{2n-1} + 1\times10^{2n-2} + ... 1\times10^0 - (2\times10^{n-1} + 2\times10^{n-2} + ... 2\times10^0)$

$= 1 + 10^1 + ... 10^{2n-1} - 2(1+10+10^2 + ...10^{n-1}) \qquad \left[\because \ S_n = \frac{a(r^n-1)}{(r-1)} \right]$

$= \frac{1(10^{2n}-1)}{9} - \frac{2(10^n-1)}{9}$

$= \frac{10^{2n} - 2.10^n - 1 + 2}{9} = \frac{10^{2n} - 2.10^n + 1}{9}$

$= \frac{(10^n-1)^2}{9} = \left(\underbrace{33......3}_{n \ time} \right)^2$ is integer.

7. $AB = 4 \Rightarrow MB = 2, \ BN = 2$

$MN = \sqrt{4+4} = \sqrt{8} = 2\sqrt{2}$

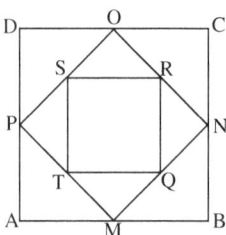

$$MN = 2\sqrt{2} \Rightarrow MQ = \sqrt{2},\ QN = \sqrt{2},\ RN = \sqrt{2}$$

$$RQ = \sqrt{QN^2 + RN^2} = 2$$

Sum of the areas $(4)^2 + \left(2\sqrt{2}\right)^2 + 2^2 + \ldots\ldots$

$$S_\infty = \frac{a}{1-r} = 32\ square\ units.$$

8. $\dfrac{a^{m+1} + b^{m+1}}{a^m + b^m} = \sqrt{ab} = a^{\frac{1}{2}} b^{\frac{1}{2}}$

$\Rightarrow \qquad a^{m+1} + b^{m+1} = a^{m+\frac{1}{2}} b^{\frac{1}{2}} + a^{\frac{1}{2}} b^{m+\frac{1}{2}}$

$\Rightarrow \qquad a^{m+\frac{1}{2}} \left(a^{\frac{1}{2}} - b^{\frac{1}{2}}\right) + b^{m+\frac{1}{2}} \left(b\frac{1}{2} - a^{\frac{1}{2}}\right) = 0$

$\Rightarrow \qquad a^{m+\frac{1}{2}} \left(a^{\frac{1}{2}} + b^{\frac{1}{2}}\right) = b^{m+\frac{1}{2}} \left(a^{\frac{1}{2}} - b^{\frac{1}{2}}\right)$

$\Rightarrow \qquad a^{m+\frac{1}{2}} = b^{m+\frac{1}{2}}$

$\Rightarrow \qquad \left(\dfrac{a}{b}\right)^{m+\frac{1}{2}} = 1 = \left(\dfrac{a}{b}\right)^0 \Rightarrow m + \dfrac{1}{2} = 0 \Rightarrow m = -\dfrac{1}{2}$

9. Let A be the first term and d be the common difference of the given A.P.

(By hypothesis)

$a = T_p = A + (p-1)d$ ... (1)

$b = T_2 = A + (q-1)d$ ... (2)

and $c = T_r = A + (r-1)d$ ... (3)

now $(1) - (2)$ and $(2) - (3)$ imply

$a - b = (p-q)d \quad and \quad b - c = (q-r)d$

$\Rightarrow \qquad \dfrac{(a-b)}{(p-q)} = \dfrac{(b-c)}{(q-r)}$

$\Rightarrow \qquad (a-b)(q-r) = (b-c)(p-q)$

$$\Rightarrow \quad a(q-r) - b(q-r) = b(p-q) - c(p-q)$$

$$\Rightarrow \quad a(q-r) + b(r-p) + c(p-q) = 0$$

10.   Let $t_1 = \log a$, $t_2 = \log b \Rightarrow d = t_2 - t_1 = \log b - \log a = \log(b/a)$

$$S_n = \frac{n}{2}(2a + (n-1)d)$$

$$= \frac{n}{2}\left(2\log a + (n-1)\log\left(\frac{b}{a}\right)\right) = \frac{n}{2}\left(\log a^2 + \log\left(\frac{b}{a}\right)^{(n-1)}\right)$$

$$= \frac{n}{2}\left(\log\frac{a^2 b^{n-1}}{a^{n-1}}\right) = \log\left(a^{3-n}\ b^{n-1}\right)^{\frac{n}{2}}$$

11.   Let a be the common ratio of G.P.

Then $1 + a + a^2 + \dots\dots + a^n = 2002$

If $n = 1$, we have $1 + 2001 = 2002$

$\therefore$ One series $\{1, 2001\}$

If $n = 2$, $1 + a + a^2$ is odd

$(\because \quad a + a^2 = a(a+1)$ is even because Product of two consecutive terms is even).

If $n = 3$, $1 + a + a^2 + a^3 = (1+a)(1+a^2)$

If a is odd, $1 + a$ & $1 + a^2$ are both even

$\Rightarrow \quad 1 + a + a^2 + a^3$ is multiple of '4'.

But 2002 is not a multiple of '4'.

If $a = 2$ *or* 4 (even) then $(1+a)(1+a^2) = odd \times odd = odd$

$\therefore$ For $n = 3$, sum of infinite series cannot be 2002.

If $n = 4$, $1 + a + a^2 + a^3 + a^4 = 1 + a(1+a) + a^3(1+a)$

$= 1 + (a+1)(a+a^3) = 1 + a(a+1)(a^2+1)$

$= 1 + even = odd \neq 2002$.

If $n \geq 5$ then $a^5 \leq 2002$

$\Rightarrow a$ may be 2, 3, 4, or 5.

If a is even then

$$1 + a + a^2 + \ldots\ldots + a^n = 1 + \underbrace{2 + 2^2 + \ldots + 2^n}_{even} = odd \neq 2002$$

Remaining possibilities are $a = 3 \; or \; a = 5$

If $a = 3$, then $S_n = \dfrac{a^{n+1} - 1}{a - 1} = \dfrac{3^{n+1} - 1}{2 - 1} = 2002$

$\Rightarrow 3^{n+1} = 4005$ is not possible $\qquad (\because 4005$ is not a multiple of 3)

If $a = 5$, then $S_n = \dfrac{5^{n+1} - 1}{5 - 1} = 2002$

$\Rightarrow 5^{n+1} = 8009$ is not possible $\qquad (\because 8009$ is not a multiple of 3)

12. $x^{18} = y^{21} = z^{28}$ $\qquad\qquad\qquad$ ... (1)

() $18 \log x = 21 \log y$

$\quad$ i.e., $\quad \log_y x = \dfrac{7}{6}$ $\qquad\qquad$ ... (2)

Now $21 \log y = 28 \log z$

i.e., $\quad \log_z y = \dfrac{4}{3}$ $\qquad\qquad$ ... (3)

$x^{18} = z^{28}$ gives $18 \log x = 28 \log z$

i.e., $\quad \log_z x = \dfrac{14}{9}$ $\qquad\qquad$ ... (4)

Now $3 \log_y x = 3 \cdot \dfrac{7}{6} = \dfrac{7}{2}$, $3 \log_z y = 3 \cdot \dfrac{4}{3} = 4$

and $7 \log_x z = 7 \cdot \dfrac{9}{14} = \dfrac{9}{2}$

i.e., $3, \dfrac{7}{2}, 4, \dfrac{9}{2}$ are in A.P.

$\therefore \; 3, 3\log_y x, 3 \log_z y, 7 \log_x z$ are in A.P.

13. $(x + a_1)(x + a_2)(x + a_3) \ldots\ldots (x + a_n)$

$$= x^n + \left(\sum a_1\right) x^{n-1} + \left(\sum a_1 a_2\right) x^{n-2} + \left(\sum a_1 a_2 a_3\right) x^{n-3} + \dots (a_1 a_2 \dots a_n)$$

$\Rightarrow$ the coefficient of x^{98} in the expansion of $(x + 1) (x + 2) (x + 3)$ .... $(x + 100)$ is sum of the products of 1, 2, 3 .... 100 taken two at a time.

Now, $(1 + 2 + 3 + \dots 100)^2 = (1^2 + 2^2 + 3^2 + \dots 100^2) + 2x$ (required sum)

$$\Rightarrow \left[\frac{100}{2}(1 + 100)\right]^2 = \frac{100(100 + 1)(200 + 1)}{6} + 2x$$

(required sum) $\Rightarrow \left(\frac{10100}{2}\right)^2 = \frac{201(10100)}{6} + 2x$ (required sum)

$\therefore$ Required Sum

$$= \frac{1}{2}\left[\left(\frac{10100}{2}\right)^2 - \frac{201(10100)}{6}\right] = 12582075$$

14. Let the number in each square be written in A.P. with a and d as first term and common difference as natural numbers.

Now Number in the last square $= a + (mn - 1) d$ and

Number in the last square in $(m - 1)$th row $= a + [m - 1) n - 1] d$
Number in the square in mth row and $(n-1)$th column

$= a + [m (n-1)] d$

Since that square are neighbours

$2[a + (mn - 1)d] = a + [(m - 1) n - 1] d + a + [m (n - 1) - 1)] d$

$\Rightarrow \qquad 2a + 2mnd - 2d = a + mnd - nd - d + a + mnd - md - d$

$\Rightarrow \qquad\qquad (m + n)d = 0$

$\Rightarrow \qquad d = 0 \qquad\qquad (\because m \neq -n)$

Hence the numbers in the square must be equal.

15. As $a_1, a_2 \dots a_n$ are in A.P. with common differnece d, so that

$$a_2 - a_1 = a_3 - a_2 = \dots = a_n - a_{n-1} = d$$

Now sin d [cosec a_1 cosec a_2 + cosec a_2 cosec a_3 + $\dots$ + cosec a_{n-1} cosec a_n]

$$= \frac{\sin(a_2 - a_1)}{\sin a_1 \sin a_2} + \frac{\sin(a_3 - a_2)}{\sin a_2 \sin a_3} + \dots + \frac{\sin(a_n - a_{n-1})}{\sin a_{n-1} \sin a_n}$$

$$= \left(\frac{\sin a_2 \cos a_2 - \cos a_2 \sin a_1}{\sin a_1 \sin a_2} \right) + \left(\frac{\sin a_3 \cos a_2 - \cos a_3 \sin a_2}{\sin a_2 \sin a_3} \right) + ..$$

$$..................... + \left(\frac{\sin a_n \cos a_{n-1} - \cos a_n \sin a_{n-1}}{\sin a_{n-1} \sin a_n} \right)$$

$$= [\cot a_1 - \cot a_2] + [\cot a_2 - \cot a_3] + ... + [\cot a_{n-1} - \cot a_n]$$

$$= \cot a_1 - \cot a_n.$$

Hence the result.

16. $\sqrt[x]{a} = \sqrt[y]{b} = \sqrt[z]{c}$

$\Rightarrow (a)^{1/x} = (b)^{1/y} = (c)^{1/z}$

$\Rightarrow \dfrac{1}{x} \log a = \dfrac{1}{y} \log b = \dfrac{1}{z} \log c$

Since, a, b, c are in G.P., we can write $b = ar$, $c = ax^2$.

$\Rightarrow \dfrac{1}{x} \log a = \dfrac{1}{y} \log (ar) = \dfrac{1}{z} \log(ar^2) = k$ (say)

$\Rightarrow x = \dfrac{\log a}{k}, y = \dfrac{\log r}{k} + , z = \dfrac{\log a}{k} + \dfrac{2 \log r}{k}$

Hence, x, y, z are in A.P. with common difference $\dfrac{\log r}{k}$

17.   Given that a, b, c are in G.P. then $b^2 = ac$

$ax^2 + 2bx + c = 0$

$\Rightarrow \qquad x = \dfrac{-2b \pm \sqrt{4b^2 - 4ac}}{2a} = \dfrac{-b}{a} \qquad [\because b^2 = ac]$

As the two equations have a common root.

$\dfrac{-b}{a}$ is a root of the second equation also.

i.e., $d \left(\dfrac{-b}{a} \right)^2 + 2e \left(\dfrac{-b}{a} \right) + f = 0$ Þ $db^2 - abe + a^2 f = 0$

$\Rightarrow \quad d\,ac - 2abe + a^2f = 0 \qquad\qquad [\because\ b^2 = ac]$

$\Rightarrow \quad \dfrac{2e}{b} = \dfrac{dc + af}{b^2} \qquad\qquad$ dividing both side by b^2

$\Rightarrow \quad \dfrac{2e}{b} = \dfrac{d}{a} + \dfrac{f}{c} \ \Rightarrow \ \dfrac{d}{a}, \dfrac{e}{b}, \dfrac{f}{c}$ are in A.P.

18.   Given $a^x = b^y = c^z = k$ (say)

     Taking logarithm, we get $x \log a = y \log b = \log c = \log k$.

$$\therefore \qquad x = \frac{\log k}{\log a}, \ y = \frac{\log k}{\log b}, \ \text{and } z = \frac{\log k}{\log c}$$

     Since x, y, z are in G.P.

$$\therefore \quad \frac{y}{x} = \frac{z}{y}$$

$$\therefore \qquad \frac{\log k}{\log b} \times \frac{\log a}{\log k} = \frac{\log k}{\log c} \times \frac{\log b}{\log k}$$

or $\qquad \dfrac{\log a}{\log b} = \dfrac{\log b}{\log c}$

i.e., $\qquad \log_b a = \log_c b$

Alternate :

Given that $a^x = b^y = c^z$

i.e., $\qquad a^x = b^y \ \& \ b^y = c^z$

$\Rightarrow \ x \log a = y \log b$ to $y \log b = z \log c$

$\Rightarrow \ \log_b^a = \dfrac{y}{x} \ \& \ \log_c^b = \dfrac{z}{y}$

Also x, y, z are in G.P. in $y^2 = 3x$

i.e., $\qquad \dfrac{y}{x} = \dfrac{z}{y}$

$\therefore \qquad \log_b^a = \log_c^b$

19.   Note that this is Arithmetico - Geometric series

(a)   Let $S = 1 + 3x + 5x^2 + 7x^3 + ... (2n - 1) x^{n-1}$

then, $xS = x + 3x^2 + 5x^3 + ... (2n - 3) x^{n-1} + (2n - 1)x^n$

$\Rightarrow S - xS = 1 + 2x + 2x^2 + 2x^3 + ... 2x^{n-1} - (2n - 1)x^n$

$\Rightarrow (1 - x) s = 1 + 2(x + x^2 + x^3 + ... x^{n-1}) - (2n - 1)x^n$

$= 1 + 2x \dfrac{(1 - x^{n-1})}{1 - x} - (2n - 1) x^n$

$\therefore S = \dfrac{1}{1 - x} + \dfrac{2x(1 - x^{n-1})}{(1 - x)^2} - \dfrac{(2n - 1)x^n}{1 - x}$

(b) $s_\infty = 1 + 3x + 5x^2 + 7x^3 + ... \Rightarrow xS_\infty = x; + 3x^2; 5x^3 + ...$

$\Rightarrow (1 - x) s_\infty = 1 + 2x + 2x^2 + 2x^3 + ...$

$= 1 + 2(x + x^2 + x^3 + ...) = 1 + \dfrac{2x}{1 - x} = \dfrac{1 + x}{1 - x}$

$\therefore \qquad S_\infty = \dfrac{1 + x}{(1 - x)^2}$

20.   Let A and H be first terms of the n AM's and n HM's respectively between the numbers a and b.

Then $A = \dfrac{an + b}{n + 1} = p$ and $H = \dfrac{ab(n + 1)}{a + bn} = q$

since n is a +ve integer, we have

$\therefore (n-1) < (n + 1) \Rightarrow 1 < \dfrac{n + 1}{n - 1}$

$\Rightarrow 1 < \left(\dfrac{n + 1}{n - 1}\right)^2$

Let q lie between p and $\left(\dfrac{n + 1}{n - 1}\right)^2$ p then $p < q$

$\Rightarrow \qquad \dfrac{ab + b}{n + 1} < \dfrac{ab(n + 1)}{a + bn}$

$\Rightarrow \qquad (an + b)(a + bn) < (n + 1)^2 ab$

$\Rightarrow \qquad (a^2 + b^2)\, n < 2abn$

$\Rightarrow \qquad (a - b)^2\, n < 0 \qquad\qquad$ But $n > 0$

$\therefore \qquad (a - b)^2 < 0$

$\therefore \qquad (a - b)^2 < 0$ which is wrong.

Hence our supposition that q lies between p and $\left(\dfrac{n+1}{n-1}\right)^2$ p is wrong

$\therefore$ q does not lie between p and $\left(\dfrac{n+1}{n-1}\right)^2$ p.

21. Series   1   4   10   22   46   .....

      Difference   3   6   12   24   .....

Let S be the sum of first n terms.

Then, $S = 1 + 4 + 10 + 22 + 46 + \dots T_n$,,,, (i)

Also, $S = 1 + 4 + 10 + 22 + \dots T_{n-1} + T_n$ ....(ii)

Subtracting (ii) from (i), we get

$0 = 1 + (3 + 6 + 12 + 24 + \dots$ to $(n-1)$ terms $- T_n \Rightarrow T_n = 1 +$

$\dfrac{3(2^{n-1} - 1)}{2 - 1} = 3.2^{n-1} - 2$

Hence, $S = \Sigma T_n = \Sigma (3.2^{n-1} - 2) = 3\,\Sigma\,2^{n-1} - \Sigma 2$

$= 3(1 + 2 + 2^2 + 2^3 + \dots 2^{n-1}) - 2n \quad = 3\dfrac{(2^n - 1)}{2 - 1} - 2n$

$= 3.2^n - 2n - 3$

22. Put $\tan^{-1}$ in the form $\tan^{-1}\left(\dfrac{x - y}{1 + xy}\right)$

Let $x - y = 2k$ and $1 + xy = 2 + k^2 + k^4$

$\Rightarrow \qquad xy = 1 + k^2 + k^4$

or $\qquad x\,(x - 2k) = 1 + k^2 + k^4$

i.e., $\qquad (x - k)^2 = (k^2 + 1)^2$

$\Rightarrow \qquad x - k = k^2 + 1$

$\Rightarrow \qquad x = k^2 + k + 1$ and Similarly $y = k^2 - k + 1$

We have $\tan^2 x - \tan^2$

$\left(\dfrac{x-y}{1+xy}\right)$ and $\dfrac{2k}{2+k^2+k^4} = \dfrac{(k^2+k+1)-(k^2-k+1)}{1+(k^2+k+1)(k^2-k+1)}$

Now $\displaystyle\sum_{k=1}^{n} \tan^{-1}\left(\dfrac{2k}{2+k^2+k^4}\right)$

$= \displaystyle\sum_{k=1}^{n} \tan^{-1}\left(\dfrac{(k^2+k+1)-(k^2-k+1)}{1+(k^2+k+1)(k^2-k+1)}\right)$

$= \displaystyle\sum_{k=1}^{n} [\tan^{-1}(k^2+k+1) - \tan^{-1}(k^2-k+1)]$

$= (\tan^{-1}3 - \tan^{-1}1) + (\tan^{-1}7 - \tan^{-1}3) + (\tan^{-1}13 - \tan^{-1}7)$

$+ \ldots\ldots\ldots + (\tan^{-1}(n^2+n+1) - \tan^{-1}.(n^2-n+1)$

$= \tan^{-1}(n^2+n+1) - \tan^{-1}1 = \tan^{-1}(n^2+n+1) - \dfrac{\pi}{4}$

23. Let $S_n = \dfrac{8}{5} + \dfrac{16}{65} + \dfrac{24}{325} + \ldots\ldots$ to n terms

$= \displaystyle\sum_{r=1}^{n}\left(\dfrac{8r}{4r^4+1}\right) = \displaystyle\sum_{r=1}^{n} \dfrac{8r}{(2r^2-2r+1)(2r^2+2r+1)}$

$= 2\displaystyle\sum_{r=1}^{n}\left(\dfrac{1}{(2r^2-2r+1)} - \dfrac{1}{(2r^2+2r+1)}\right)$

$= 2\left[\dfrac{1}{1} - \dfrac{1}{5} + \dfrac{1}{5} - \dfrac{1}{13} + \dfrac{1}{13} - \ldots + \dfrac{1}{(2n^2-2n+1)} - \dfrac{1}{(2n^2+2n+1)}\right]$

$= 2\left[1 - \dfrac{1}{2n^2+2n+1}\right] = \dfrac{4n^2+4n}{2n^2+2n+1}$

24. $x = 1 + 3a + 6a^2 + 10a^3 \ldots\ldots\ldots \quad ax = a + 3a^2 + 6a^3$

$\therefore \quad x(1-a) = 1 + 2a + 3a^2 + 4a^3 \ldots\ldots\ldots$

$\Rightarrow \quad xa(1-a) = a + 2a^2 + 3a^3 \ldots\ldots\ldots$

$\therefore \quad x(1-a) - x(1-a)a = 1 + a + a^2 + a^3 \ldots\ldots$

$$\Rightarrow \quad x(1-a)^2 = \frac{1}{1-a} \quad \Rightarrow \quad x = \frac{1}{(1-a)^3}$$

$$\therefore \quad a = 1 - x^{-1/3}$$

Similarly $b = 1 - y^{-1/4}$

Now $s = 1 + 3ab + 5(ab)^2$ ....

$abs = ab + 3\,(ab)^2$ ....

$$\Rightarrow \quad s(1-ab) = 1 + 2ab + 2(ab)^2$$

$$= 1 + \frac{2ab}{(1-ab)} = \frac{1+ab}{1-ab}$$

$$s = \frac{1+ab}{(1-ab)^2}$$

Put the values of a and b.

$$\Rightarrow \quad s = \frac{1+(1-x^{-(1/3)})(1-y^{-(1/4)})}{[1-(1-x^{-(1/3)})(1-y^{-(1/4)})]^2}$$

$$s = \frac{[x^3y^3+(x^3-1)(y^3-1)]x^3y^3}{[x^3y^3-(x^3-1)(y^3-1)]^2} \cdot$$

25. For three positive numbers x, y, z, we have

$$A = \frac{x+y+z}{3}, \; H = \frac{1}{\dfrac{\dfrac{1}{x}+\dfrac{1}{y}+\dfrac{1}{z}}{3}} \quad \text{and } A > H$$

Here, $s - a$, $s - b$, $s - c$ are positive

$$\therefore \quad \frac{s-a+s-b+s-c}{3} > \frac{1}{\dfrac{\dfrac{1}{s-a}+\dfrac{1}{s-b}+\dfrac{1}{s-c}}{3}}$$

$$\Rightarrow \quad \frac{3}{2s} < \frac{\dfrac{1}{s-a}+\dfrac{1}{s-b}+\dfrac{1}{s-c}}{3}$$

$$\Rightarrow \frac{3s-(a+b+c)}{3} > \frac{1}{\dfrac{1}{s-a}+\dfrac{1}{s-b}+\dfrac{1}{s-c}}$$

$$\therefore \frac{1}{s-a}+\frac{1}{s-b}+\frac{1}{s-c} > \frac{9}{2s}$$

26. $f(x) = 7 + 2x\ ln\ 25 - 5^{x-1} - 5^{2-x} = 7 + 2x\ ln25 - \dfrac{5^x}{5} - \dfrac{25}{5^x}$

$\therefore f'(x) = 4\ ln\ 5 - \dfrac{5^x}{5}\ ln5 + 25.5^{-x}\ ln5$

$= \dfrac{ln5}{5}\ (20 - 5^x + 125.5^{-x})\ (5^x - 25)$

$= \dfrac{ln5}{5} \left[\dfrac{5^{2x} - 20.5^x - 125}{5^x} \right] = \dfrac{-ln5}{5^{x+1}}\ /\ (5^x + 5)$

for $f'(x) = 0 \Rightarrow 5^x - 5^2 = 0 \Rightarrow x = 2$ but $5^x \neq -5$

now $f''(x)$ at $x = 2 < 0 \Rightarrow$ maximum at $x = 2$

$\therefore a = 2$

also $r = \displaystyle\lim_{x\to 0} \int_0^x \frac{t^2 dt}{(x^2 \tan(\pi + x))} = \lim_{x\to 0} \frac{x^3}{3x^2 \tan(\pi + x)} = \frac{1}{3}$

$\therefore$ Sum $= a + ar + ar^2 \ldots\ldots\ldots\ldots = a\ (1 + r + r^2\ldots\ldots\ldots\infty)$

$= a\left(\dfrac{1}{1-r}\right) = \dfrac{2}{(2/3)} = 3.$

27.   Let k^{th} term of I^{st} A.P., $t_k = 17 + (k-1)\ 4$
      $= 21 + (k - 2)\ 4$ and r^{th} term of 2nd A.P.
      $t'_r = 16 + (r - 1)\ 5 = 21 + (r - 2)\ 5$
      $\therefore$ The common terms $4\ (k - 2) = 5(r - 2)$

$\Rightarrow \qquad r = \left(\dfrac{2 + 4k}{5}\right)$

First term of new series $t_1 = 21$ this happens for $k = 2$ and $r = 2$

193

Second term will be when $k - 2 = 5$ and $r - 2 = 4$

i.e., $k = 7$ & $r = 6$

$\Rightarrow t_2 = 17 + (7 - 1) \times 4 = 41$

$\therefore$ Common difference $= 41 - 21 = 20$. and there are 100 terms.

$\therefore \ \text{sum} = \dfrac{100}{2} [2 \times 21 + (100 - 1) \times 20]$

$\quad = 101100$.

Note : Recoming decimals furnish a good illustration of inifinite Geometrical progressions.

28.   Here, $T_n = n(n + 1)(n + 2) = n^3 + 3n^2 + 2n$

$\quad s_n = \Sigma T_n = \Sigma (n^3 + 3n^2 + 2n)$

$\qquad = \Sigma n^3 + 3\Sigma n^2 + 2\Sigma n$

$= \dfrac{n^2(n + 1)^2}{4} + \dfrac{3n(n + 1)(2n + 1)}{6} + \dfrac{2n(n + 1)}{2}$

$= \dfrac{n(n + 1)}{4} [n(n+1)+2(2n+1)+4]$

$= \dfrac{n(n + 1)}{4} [n^2 + 5n + 6] = \dfrac{n(n + 1)(n + 2)(n + 3)}{4}$

29.   Let the three numbers be $\dfrac{a}{r}$, a, ar

$\therefore \left(\dfrac{a}{r}\right)^2 + a^2 + (ar)^2 = s^2$ and $\dfrac{a}{r} + a + ar = \alpha$, s

Let $r + \dfrac{1}{r} = t \Rightarrow r^2 + \dfrac{1}{r^2} = t^2 - 2$

$\therefore t^2 - 2 + 1 = \dfrac{s^2}{r^2}$ or $t^2 - 1 = \left(\dfrac{s}{a}\right)^2$ ... (1)

and $t + 1 = \dfrac{\alpha.s}{a} \Rightarrow \left(\dfrac{s}{a}\right)^2 = \left(\dfrac{t + 1}{\alpha}\right)^2 = \dfrac{t^2 + 1 + 2t}{\alpha^2}$ ... (2)

From 1st and 2nd

$$\therefore \quad \frac{t^2 + 2t + 1}{\alpha^2} = t^2 - 1 \Rightarrow t^2 + 2t + 1 = \alpha^2 t^2 - \alpha^2$$

or $\quad \alpha^2 = \dfrac{(t+1)^2}{(t^2-1)} = \dfrac{(t+1)}{(t-1)} \Rightarrow t = \dfrac{\alpha^2+1}{\alpha^2-1} = r + \dfrac{1}{r}$

$\Rightarrow \quad ((\alpha^2 - 1))\, r^2 - (\alpha^2 + 1)\, r + (\alpha^2 - 1) = 0$

as r is real D $\geq$ 0

or $\quad (\alpha^2 + 1)^2 - 4(\alpha^2 - 1)^2 \geq 0$

or $\quad (3\alpha^2 - 1)(3 - \alpha^2) \geq 0 \Rightarrow \alpha^2 \geq \dfrac{1}{3} \ and \ \alpha^2 \geq 3$

Also at $\alpha^2 = \dfrac{t+1}{t-1}$, $\alpha^2 = \dfrac{1}{3}$ or 3 i.e., t = 2 or –2 which is not true for

r $\neq$ 1, –1.

Also at $\alpha^2 = 1 \ t + 1 \neq t - 1$

$$\alpha^2 \in \left[\frac{1}{3}, 1\right] \cup (1, 3)$$

30. Remember,

$123 = 100 + 20 + 3$

or xyz = 100 x + 10y + 1z

Now, let's take three numbers in G.P. be a, ar, ar^2

Given, $a(ar)(ar^2) - 792 = (ar^2)(ar)(a)$ .... (1)

Also, given a, (ar + 2), ar^2 are in A.P.

$\Rightarrow 2(ar + 2) = a + ar^2$

$\Rightarrow ar^2 - 2ar + a = 4$

$\Rightarrow a(r-1)^2 = 4$ ..... (2)

From equation (1)

$100a + 10ar + ar^2 - 792 = 100ar^2 + 10\,ar + a$

$\Rightarrow 99ar^2 - 99a + 792 = 0$

$\Rightarrow 99a\,(r^2 - 1) = -792$

$\Rightarrow a\,(r^2 - 1) = -8$ .... (3)

195

Dividing (3) by (2)

$$\frac{(r+1)}{r-1} = -2$$

$\Rightarrow r + 1 = -2r + 2$

$\Rightarrow 3r = 1$

$$r = \frac{1}{3}$$

$a = 9$

Hence the numbers are 9, 3, 1 i.e. 931

31. $a_1 + a_5 + a_{10} + a_{15} + a_{20} + a_{24} = 225$

$\Rightarrow (a_1 + a_{24}) + (a_5 + a_{20}) + (a_{10} + a_{15}) = 225$

$\Rightarrow 3(a_1 + a_{24}) = 225 \Rightarrow a_1 + a_{24} = 75$

Hence, $S_{24} = \dfrac{24}{2} (a_1 + a_{24}) = 12 \,(75) = 900$

32. Let 'a' and 'd' be the first term and common difference respectively.

Then, $\dfrac{S_m}{S_n} = \dfrac{\dfrac{m}{2}[2a + (m-1)d]}{\dfrac{n}{2}[2a + (n-1)d]} = \dfrac{m^2}{n^2}$

$\Rightarrow 2an + (m-1)\,nd = 2am + (n-1)\,md$

$\Rightarrow 2a(n-m) = (n-m)d \Rightarrow d = 2a$

Now, $\dfrac{T_m}{T_n} = \dfrac{a + (m-1)d}{a + (n-1)d} = \dfrac{a + (m-1)(2a)}{a + (n-1)(2a)}$

$= \dfrac{2m-1}{2n-1}$

33. $5 + 55 + 555 \ldots\ldots$ to n terms

$= \dfrac{5}{9}\,(9 + 99 + 999 + \ldots\ldots \text{ to n terms})$

$= \dfrac{5}{9}\,[(10-1) + (100-1) + (1000-1) + \ldots (10^n-1)]$

$$= \frac{5}{9} [(10 + 10^2 + 10^3 + \dots 10^n) - n]$$

$$= \frac{5}{9} = \left[\frac{10(10^n - 1)}{10 - 1} - n \right] = \frac{5}{81} \{10^{n+1} - 10 - 9n\}$$

34. a, b, c are in A.P., $\Rightarrow 2b = a + c$ ... (i)

α, β, γ are in H.P. $\Rightarrow \beta = \dfrac{2\alpha\gamma}{\alpha + \gamma}$ ... (ii)

$a\alpha, b\beta, c\gamma$ are in G.P. $\Rightarrow b^2\beta^2 = ac\alpha\gamma$ ... (iii)

From (i), (ii) and (iii) $= \left(\dfrac{a + c}{2} \right)^2 \left(\dfrac{2\alpha\gamma}{\alpha + \gamma} \right)^2 = ac\alpha\gamma$

$\Rightarrow \left(\dfrac{a + c}{ac} \right)^2 = \dfrac{(\alpha + \gamma)^2}{\alpha\gamma} \Rightarrow \dfrac{a}{c} + \dfrac{c}{a} + 2 = \dfrac{\alpha}{\gamma} + \dfrac{\gamma}{\alpha} + 2$

$\Rightarrow \dfrac{\alpha}{\gamma} \left(\dfrac{a}{c} + \dfrac{c}{a} \right) = \dfrac{\alpha}{\gamma} \left(\dfrac{\alpha}{\gamma} + \dfrac{\gamma}{\alpha} \right) \Rightarrow \dfrac{\alpha^2}{\gamma^2} - \dfrac{\alpha}{\gamma} \left(\dfrac{a}{c} + \dfrac{c}{a} \right) + 1 = 0$

$\Rightarrow \dfrac{\alpha^2}{\gamma^2} - \dfrac{\alpha}{\gamma} \cdot \dfrac{a}{c} - \dfrac{c}{a} \cdot \dfrac{\alpha}{\gamma} + \dfrac{c}{a} \cdot \dfrac{a}{c} = 0$

$\Rightarrow \dfrac{\alpha}{\gamma} \left(\dfrac{\alpha}{\gamma} - \dfrac{a}{c} \right) - \dfrac{c}{a} \left(\dfrac{\alpha}{\gamma} - \dfrac{a}{c} \right) = 0$

$\Rightarrow \left(\dfrac{\alpha}{\gamma} - \dfrac{c}{a} \right) \left(\dfrac{\alpha}{\gamma} - \dfrac{a}{c} \right) = 0$

$\Rightarrow \dfrac{\alpha}{\gamma} = \dfrac{c}{a}$ or $\dfrac{\alpha}{\gamma} = \dfrac{a}{c} \Rightarrow a\alpha = \gamma c$ or $\alpha c = a\gamma$

Since $a\alpha, b\beta, \gamma c$ are in G.P.,

$a\alpha \neq \gamma c$

$\Rightarrow \qquad \alpha c = a\gamma$ ... (iv)

197

From (iii) and (iv) $\alpha c = b\beta = a\gamma$

$$\Rightarrow \qquad \frac{c}{1/\alpha} = \frac{b}{1/\beta} = \frac{\alpha}{1/\gamma}$$

Hence $a : b : c = \dfrac{1}{\gamma} : \dfrac{1}{\beta} : \dfrac{1}{\alpha}$

35.   Coefficient of x^{n-1} in $(1 + 2x + 3x^2 \dots nx^{n-1})^2$

$$= 1.n + 2.(n{-}1) + 3.(n{-}2) \dots n.1$$

$$= \sum_{r=1}^{n} r(n - r + 1) = \sum_{r=1}^{n} (nr - r^2 + r)$$

$$= \frac{(n+1)n(n+1)}{2} - \frac{n(n+1)(2n+1)}{6}$$

$$= \frac{n(n+1)(3n + 3 - (2n+1))}{6}$$

$$= \frac{n(n+1)(n+2)}{6}.$$

36. $x = 1 + a + a^2 + \dots\dots\dots\dots\dots\dots + \infty$

$$= \frac{1}{1-a}, \; [\because \; |a| < 1]$$

$$\Rightarrow \qquad a = \frac{x-1}{x};$$

$$y = 1 + b + b^2 + \dots\dots\dots\dots + \infty \; = \frac{1}{1-b}$$

$$\Rightarrow \qquad b = \frac{y-1}{y}$$

$$\therefore \qquad 1 + ab + a^2b^2 + \dots\dots\dots\dots + \infty$$

$$= \frac{1}{1-ab} = \frac{1}{1 - \dfrac{(x-1)(y-1)}{xy}} = \frac{xy}{x+y-1}$$

1. Find the sum of the series $1017 + 1035 + 1053 + \ldots\ldots\ldots + 9999$

2. Find the arithmetic progression if $a_2 - a_6 + a_4 = -7$, $a_8 - a_7 = 2a_4$

3. How many number of two digits are divisible by 7?

4. Find (1) 'T_n' and (2) S_n of the series $1 + \dfrac{1}{3} + \dfrac{1}{3^2} + \dfrac{1}{3^3} + \ldots\ldots\ldots$

5. Find the sum of n terms of the geometric series $243 + 324 + 432 + \ldots\ldots\ldots$

6. Evaluate $\displaystyle\sum_{n=2}^{7}\left(2^n + 3^{n-1}\right)$

7. How many terms of the geometric sequence $\sqrt{3},\ 3,\ 3\sqrt{3},\ \ldots\ldots\ldots$ must be taken to make the sum $39 + 13\sqrt{3}$?

8. Between what limits r must lie in order that the series
$$\left(\frac{2r}{r+3}\right) + \left(\frac{2r}{r+3}\right)^2 + \left(\frac{2r}{r+3}\right)^3 + \ldots\ldots\ldots \text{ may have a definite sum?}$$

9. If $a-b,\ b-c,\ c-a$ are in G.P. show that $(a+b+c)^2 = 3(ab+bc+ca)$

EXERCISE - 2
Answers

1. $27,54,000$

2. $\{-5, 3, 1, 3, 5, \ldots\ldots\}$

3. 13

4. $\left(\dfrac{1}{3}\right)^{n-1}$; $\dfrac{3}{2}\left(1 - \dfrac{1}{3^n}\right)$

5. $3^{6-n}(4^n - 3^n)$

6. 1344

7. 6

8. $1 < r < 3$

blank page

1. ALGEBRAIC IDENTITY

EXERCISE - 2
SOLUTIONS

1. Suppose there are n pages in the novel. Since the sum of the numbers on the pages must exceed 15000, therefore

$$\frac{n(n+1)}{2} > 15000$$

i.e., $\qquad n(n+1) > 30000$

i.e., $\qquad (n+1)^2 > n(n+1) > 30000 > 173^2$

which gives $n > 172$.

Also, the sum of the numbers on the torn leaf $\leq 2n - 1$.

Therefore $(1 + 2 + ... + n) - (2n - 1) \leq 15000$,

i.e., $\qquad \dfrac{(n-1)(n-1)}{2} \leq 15000$

Now, $(n-2)^2 < (n-2)(n-1) \leq 30000 < 174^2$

Therefore $(n-2)^2 < 174^2$, i.e., $n < 176$

Thus we find that $172 < n < 176$

The possible values of n are therefore $173, 174, 175$.

Case 1. Let $n = 173$

Then $\dfrac{n(n+1)}{2} = 15051$,

showing that the page numbers on the torn leaf are together equal to 51. If the page numbers are x, $x+1$, we have

$2x + 1 = 51$, which gives $x = 25$.

The page numbers on the torn leaf are thereofre 25, 26.

Case 2. Let $n = 174$.

Then $\dfrac{n(n+1)}{2} = 15225$,

so that the page numbers on the torn leaf are together equal to 225.

$2x + 1 = 225$, gives $x = 112$.

The page numbers on the torn leaf are 112, 113. While theoretically this is possible, but in practice, the smaller number on a leaf is odd, therefore we reject these values.

Case 3. Let $n = 175$. Then,

$$\frac{n(n+1)}{2} = \frac{175.176}{2} = 15400.$$

The sum of the page numbers on the torn leaf = 400, which is not possible since the maximum value of the sum of the numbers on the torn leaf can be 173 + 174 = 347, and so this value of n has to be rejected.

Hence the page numbers on the torn leaf are 25, 26.

2.   See solution no. 10 in exercise 1.

4.   Suppose the given number is $a_n a_{n-1} a_{n-2} \ldots\ldots a_1 7$.

i.e., $a_n.10^n + a_{n-1}.10^{n-1} + \ldots\ldots + 10a_1 + 7$

i.e., $10x + 7$, where $x = a_n 10^{n-1} + a_{n-1} 10^{n-2} + \ldots\ldots + a_1$.

When the last digit is carried to the beginning of the number, it becomes $7a_n a_{n-1} \ldots\ldots a_1$,

i.e., $7.10^n + a_n 10^{n-1} + a_{n-2} 10^{n-2} + \ldots\ldots + a_1$,

i.e., $7.10^n + x$.

We are given that $7.10^n + x = 5(10x + 7)$

so that $7x = 10^n - 5$ ... (i)

We shall find the least value of n, and consequently that of x as well, for which (i) has an integer solution for x.

The smallest value of n for which 10^n leaves a remainder 5 when divided by 7 turns out to be 5 (by actually dividing 1000 ..... 0 by 7. In fact $100000 = (14285) \times 7 + 5$.

The smallest number is then 142857.

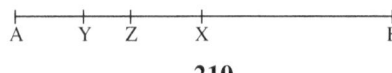

210

5. If $x = 5 + 2\sqrt{6} = \left(\sqrt{3} + \sqrt{2}\right)^2$, then

$$\sqrt{x} = \sqrt{3} + \sqrt{2}$$

The value of $\sqrt{x} + 1/\sqrt{x} = \left(\sqrt{3} + \sqrt{2}\right) + \dfrac{1}{\left(\sqrt{3} + \sqrt{2}\right)}$,

$$= \left(\sqrt{3} + \sqrt{2}\right) + \left(\sqrt{3} - \sqrt{2}\right),$$

$$= 2\sqrt{3}$$

Also, $\sqrt{x} - 1/\sqrt{x} = \left(\sqrt{3} + \sqrt{2}\right) = \dfrac{1}{\left(\sqrt{3} + \sqrt{2}\right)}$

$$= \left(\sqrt{3} + \sqrt{2}\right) - \left(\sqrt{3} - \sqrt{2}\right) = 2\sqrt{2}$$

9. Let $y - z = a$, $z - x = b$, and $x - y = c$.

Then, $a + b + c = 0$ and consequently

$$a^3 + b^3 + c^3 = 3abc,$$

$$(y - z)^3 = (z - x)^3 + (x - y)^3 = 3(y - z)(z - x)(x - y).$$

3. QUADRATIC EQUATION

EXERCISE - 2
SOLUTIONS

1. The equation is $abx^2 - (a^2 + b^2)x + ab = 0$

$$x = \frac{(a^2 + b^2) \pm \sqrt{(a^2 + b^2)^2 - 4a^2b^2}}{2ab}$$

$$= \frac{(a^2 + b^2) \pm \sqrt{(a^2 - b^2)^2}}{2ab}$$

$$= \frac{(a^2 + b^2) \pm (a^2 - b^2)}{2ab}$$

$$= \frac{a^2 + b^2 + a^2 - b^2}{2ab} \quad or \quad \frac{a^2 + b^2 - a^2 + b^2}{2ab}$$

$$= \frac{a}{b} \quad or \quad \frac{b}{a}$$

2. We know that

$$\left(x + \frac{1}{x}\right)^2 = x^2 + \frac{1}{x^2} + 2$$

$$\Rightarrow \quad x^2 + \frac{1}{x^2} = \left(x + \frac{1}{x}\right)^2 - 2$$

The given equation becomes

$$16\left[\left(x + \frac{1}{x}\right)^2 - 2\right] = 257$$

$$\Rightarrow \quad \left(x + \frac{1}{x}\right)^2 - 2 = \frac{257}{16}$$

$$\Rightarrow \quad \left(x + \frac{1}{x}\right)^2 = \frac{289}{16}$$

$$x + \frac{1}{x} = -\frac{17}{4} \quad \text{or} \quad x + \frac{1}{x} = -\frac{17}{4}$$

$$\Rightarrow \quad x + \frac{1}{x} = -\frac{17}{4} \quad \text{or} \quad \frac{x^2 + 1}{x} = -\frac{17}{4}$$

$$4x^2 - 17x + 4 = 0 \quad \text{or} \quad 4x^2 + 17x + 4 = 0$$

$$\Rightarrow \quad x = \frac{17 \pm \sqrt{289 - 64}}{8} \quad \text{or} \quad x = \frac{-17 \pm \sqrt{289 - 64}}{8}$$

$$\Rightarrow \quad x = \frac{17 \pm 15}{8} \quad \text{or} \quad x = \frac{-17 \pm 15}{8}$$

$$\Rightarrow \quad x = 4 \ or \ \frac{1}{4} \quad \text{or} \quad x = -4 \ or \ -\frac{1}{4}$$

3.
$$(a+b)\left(x + \sqrt{1+x^2}\right) = 2a\sqrt{1+x^2}$$

$$(a+b)x + a\sqrt{1+x^2} + b\sqrt{1+x^2} = 2a\sqrt{1+x^2}$$

$$(a+b)x = (a-b)\sqrt{1+x^2} \qquad \ldots (1)$$

Squaring both sides,

$$(a+b)^2 x^2 = (a-b)^2 (1+x^2)$$

$$x^2\left[(a+b)^2 - (a-b)^2\right] = (a-b)^2$$

$$x^2[4ab] = (a-b)^2$$

$$\Rightarrow \quad x^2 = \frac{(a-b)^2}{4ab}$$

$$x = \pm \frac{a-b}{2\sqrt{ab}}$$

Note: Equation (1) is $(a+b)x = \sqrt{1+x^2} \ (a-b)$

Here $\sqrt{}$ always means positive root. Thus the solution $x = \dfrac{a-b}{2\sqrt{ab}}$

213

satisfies the equation. But not $x = \dfrac{-(a-b)}{2\sqrt{ab}}$. This solution arises because we had squared equation (1). This solution is called the extraneous root. The student will become clear when a numerical example is taken.

Consider the equation $\sqrt{x} = x - 2$

Squaring (1) we get,

$$x = (x-2)^3$$

$$x = x^2 - 4x + 4$$

$$x^2 - 5x + 4 = 0$$

$$(x-4)(x-1) = 0$$

$$x = 1, 4$$

When we put $x = 1$ in (1) we get $1 = 1 - 2 = -1$. Put $x = 4$, $\sqrt{4} = 4 - 2$, $2 = 2$ it satisfies.

$\therefore$ $x = 4$ is the solution, $x = 1$ is the extraneous root.

4. The given equation can be written as

$$\frac{3^x}{3^2} + \frac{3^3}{3^x} = 4$$

Put $y = 3^x$, then $\dfrac{y}{9} + \dfrac{27}{y} = 4$

$$\Rightarrow \qquad \frac{y^2 + 243}{9y} = 4$$

$$\Rightarrow \qquad y^2 + 243 = 36y$$

$$\Rightarrow \qquad y^2 - 36y + 243 = 0$$

$$\Rightarrow \qquad y^2 - 9y - 27y + 243 = 0$$

$$\Rightarrow \qquad y(y-9) - 27(y-9) = 0$$

$$\Rightarrow \qquad (y-9)(y-27) = 0$$

$$\Rightarrow \quad y = 9 \quad or \quad y = 27$$

$$\Rightarrow \quad 3^x = 9 \quad or \quad 3^x = 27$$

$$3^x = 3^2 \quad or \quad 3^x = 3^3$$

$$\Rightarrow \quad x = 2 \ or \ x = 3$$

5. The equation can be written as

$$(x+2)(x+5)(x+3)(x+4) = 24(x^2 + 7x + 7)$$

$$(x^2 + 7x + 10)(x^2 + 7x + 12) = 24(x^2 + 7x + 7)$$

Put $\quad x^2 + 7x = y$

$$(y+10)(y+12) = 24(y+7)$$

$$\Rightarrow \quad y^2 - 2y - 48 = 0$$

$$\Rightarrow \quad (y-8)(y+6) = 0$$

$$\Rightarrow \quad y = 8, -6$$

when $y = 8$ we have $x^2 + 7x = 8 \Rightarrow x^2 + 7x - 8 = 0, x = 1, -8$

When $y = -6$ we have

$$x^2 + 7x = -6$$

$$x^2 + 7x + 6 = 0$$

$$(x+1)(x+6) = 0$$

$$x = -1 \quad or \quad -6$$

Solution set $\{-1, 1, -6, -8\}$.

6. We have $\left(x^2 - 3x + 36\right) - \left(x^2 - 3x + 9\right) = 27$.

Taking this as

$$\left(\sqrt{x^2 - 3x + 36}\right)^2 - \left(\sqrt{x^2 - 3x + 9}\right)^2 = 27$$

$$\left(\sqrt{x^2 - 3x + 36} + \sqrt{x^2 - 3x + 9}\right)\left(\sqrt{x^2 - 3x + 36} - \sqrt{x^2 - 3x + 9}\right) = 27$$

Using the given equation we get,

$$\sqrt{x^2-3x+36}+\sqrt{x^2-3x+9}=0 \qquad \ldots (2)$$

From (1) and (2), we get

$$2\sqrt{x^2-3x+36}=12$$

$$\Rightarrow \qquad \sqrt{x^2-3x+36}=6$$

$$x^2-3x+36=36$$

$$\Rightarrow \qquad x^2-3x=0$$

$$x(x-3)=0$$

$$\Rightarrow \qquad x=0 \quad or \quad x=3$$

7. Squaring both sides, we get

$$x+5+x+12+2\sqrt{x+5}\sqrt{x+12}=2x+41$$

$$\Rightarrow \qquad \sqrt{(x+5)(x+12)}=12$$

$$\Rightarrow \qquad (x+5)(x+12)=144$$

$$\Rightarrow \qquad x^2+17x-84=0$$

$$\Rightarrow \qquad (x-4)(x+21)=0$$

$$x=4 \qquad or \qquad -21$$

$x=4$ satisfies the equation and $x=-21$ is not a solution.

∴ Solution is $x=4$.

8. Dividing by x^2 (since $x=0$ does not satisfy the equation)

we have, $\qquad 3x^2-20x-94-\dfrac{20}{x}+\dfrac{3}{x^2}=0$

$$3\left(x^2+\dfrac{1}{x^2}\right)-20\left(x+\dfrac{1}{x}\right)-94=0 \qquad \ldots (1)$$

Put $\qquad x+\dfrac{1}{x}=y \qquad \Rightarrow \qquad x^2+\dfrac{1}{x^2}=y^2-2$

216

(1) $\Rightarrow 3y^2 - 20y - 100 = 0 \Rightarrow y = -\dfrac{10}{3}$ *or* 10

$$y = -\frac{10}{3} \qquad \Rightarrow \qquad x + \frac{1}{x} = -\frac{10}{3}$$

$$\text{or} \qquad x = -3 \ or \ -\frac{1}{3}$$

$$y = 10 \quad \Rightarrow \qquad x + \frac{1}{x} = 10 \Rightarrow x = 5 \pm \sqrt{24}$$

Roots are $-\dfrac{1}{3}, \ -3, \ 5 + \sqrt{24}$ *and* $5 - \sqrt{24}$.

4. SYSTEMS OF ALGEBRAIC EQUATIONS

EXERCISE - 2
SOLUTIONS

1. We have

$$x + y = 3 \qquad \ldots (1)$$

$$\frac{x^2}{y} + \frac{y^2}{x} = \frac{9}{2} \qquad \ldots (2)$$

From (1) $y = 3 - x$. Substituting (2) we get,

$$\frac{x^2}{3-x} + \frac{(3-x)^2}{x} = \frac{9}{2} \Rightarrow x^2 - 3x + 2 = 0 \Rightarrow x = 1, 2 .$$

When $x = 1$, we get $y = 2$

When $x = 2$, we get $y = 1$

$\therefore$ The solution set are $(1, 2)$, $(2, 1)$

2. We have

$$3x^2 - 3mx^2 + m^2x^2 = 7 \qquad \ldots (1)$$

and $\quad 2x^2 - 3mx^2 + 2m^2x^2 = 14 \qquad \ldots (2)$

$$\frac{(1)}{(2)} \quad \Rightarrow \quad \frac{(3 - 3m + m^2)}{(2 - 3m + 2m^2)} = \frac{1}{2}$$

Simplyfing we get $m = \dfrac{4}{3}$

$$(1) \quad \Rightarrow \quad 3x^2 - 3, \frac{4}{3}x^2 + \frac{16}{9}x^2 = 7$$

$$x^2 = 9$$

$$\Rightarrow \quad x = \pm 3$$

When $x = 3$ we have $y = \dfrac{4}{3}x$

$$\therefore \qquad y = 4$$

When $x = -3$ we have $y = -4$

The solution sets are $(3, 4)$ and $(-3, -4)$

3.
$$x + y = 2 \qquad \qquad \dots (1)$$

$$xy - z^2 = 1 \qquad \qquad \dots (2)$$

$(2) \quad \Rightarrow \quad xy = 1 + z^2$

Now, $(x - y)^2 = (x + y)^2 - 4xy$

$(x - y)^2 = 4 - 4(1 + z^2) = -4z^2$

$\Rightarrow \quad x - y = \sqrt{-4z^2}$

Since x and y are real and since z^2 is positive.

$-4z^2$ is negative

$\therefore \quad \sqrt{-4z^2}$ is imaginary.

$\Rightarrow \quad x - y = 0$ and $x + y = 2$

$\Rightarrow \quad x = 1, \ y = 1$

4. Let $x + y = u, \dfrac{x}{y} = v$

$\therefore \quad u + v = \dfrac{1}{2}$

$uv = -\dfrac{1}{2}$

$(u - v)^2 = (u + v)^2 - 4uv$

$$= \frac{1}{4} - 4\left(-\frac{1}{2}\right)$$

$$= \frac{1}{4} + 2 = \frac{9}{4}$$

$$\therefore \qquad u-v=\pm\frac{3}{2},\ u-v=\frac{3}{2} \Rightarrow u=1,\ v=-\frac{1}{2}$$

$$\therefore \qquad u+v=\frac{1}{2},\ u-v=-\frac{3}{2} \Rightarrow u=-\frac{1}{2},\ v=1$$

Case (1):

$$x+y=1,\ \frac{x}{y}=-\frac{1}{2} \Rightarrow x=-1,\ y=2$$

Case (2):

$$x+y=-\frac{1}{2},\ \frac{x}{y}=1 \Rightarrow x=-\frac{1}{4},\ y=-\frac{1}{4}$$

since $x<0,\ y<0$, the equation set is $\left(-\frac{1}{4},\ -\frac{1}{4}\right)$

5.   We have

$$x^4+y^4=82 \qquad\qquad \ldots (1)$$

$$x+y=4 \qquad\qquad \ldots (2)$$

Putting $\quad x=u+v,\ y=u-v$ we get

From (2) $u+v+u-v=4$

$$\Rightarrow \qquad u=2$$

(1) $\qquad \Rightarrow \qquad (u+v)^4+(u-v)^4=82$

$$\Rightarrow \qquad (2+v)^4+(2-v)^4=82$$

expanding and simplyfing we get

$$v^4+24v^2-25=0$$

$$\Rightarrow \qquad (v^2-1)(v^2+25)=0$$

$$v^2=1 \quad \Rightarrow \quad v=\pm1$$

$$v^2+25=0 \qquad \Rightarrow \qquad v \text{ is imaginary.}$$

The solution set can be got from $x=u+v,\ y=u-v$.

$u=2,\ v=1 \qquad \Rightarrow \qquad x=3,\ y=1$

220

$$u = 2, v = -1 \qquad\qquad x = 1, y = 3$$

$$v^2 + 25 = 0 \qquad \Rightarrow \qquad v = \pm 5i$$

$$\Rightarrow \qquad x = 2 - 5i$$

$$y = 2 + 5i$$

6. The first two equations can be written as

$$x - cy - bz = 0 \qquad\qquad\qquad \text{... (1)}$$

$$-cx + y - az = 0 \qquad\qquad\qquad \text{... (2)}$$

Solve these two equations using cross multiplication rule

$$\frac{x}{-ac - b} = \frac{y}{-bc - a} = \frac{z}{-1 + c^2} = K \ (say)$$

$$x \qquad = \qquad K(-ac - b)$$

$$y \qquad = \qquad K(-bc - a)$$

$$z \qquad = \qquad K(-1 + c^2)$$

Substitute these values in the third equation given $z = bx + ay$

$$K(-1 + c^2) \qquad = \qquad bK(-ac - b) + aK(-bc - a)$$

$$c^2 - 1 \quad = \qquad -b(b + ac) - a(a + bc)$$

$$c^2 - 1 \quad = \qquad -b^2 - abc - a^2 - abc$$

$$\Rightarrow \qquad\qquad a^2 + b^2 + c^2 + 2abc = 1$$

7.
$$x^2 - yz = a^2 \qquad\qquad \text{... (1)}$$

$$y^2 - az = b^2 \qquad\qquad \text{... (2)}$$

$$z^2 - xy = c^2 \qquad\qquad \text{... (3)}$$

Multiply (1) by y, (2) by z and (3) by x. We get

$$x^2 y - y^2 z \qquad = \qquad a^2 y$$

$$y^2 z - z^2 x \qquad = \qquad b^2 z$$

$$z^2 x - x^2 y \qquad = \qquad c^2 x$$

Add all,

$$a^2 y + b^2 z + c^2 x = 0 \qquad \ldots (4)$$

Multiply (1) by z, (2) by x and (3) by y

$$x^2 z - yz^2 \quad = \quad a^2 z$$

$$y^2 x - zx^2 \quad = \quad b^2 x$$

$$z^2 y - xy^2 \quad = \quad c^2 y$$

Add all,

$$a^2 z + b^2 x + c^2 y = 0 \qquad \ldots (5)$$

We write the equations (4) and (5) again below

$$c^2 x + a^2 y + b^2 z = 0$$

$$b^2 x + c^2 y + a^2 z = 0$$

Using the rule of cross multiplication we get

$$\frac{x}{a^4 - b^2 c^2} = \frac{y}{b^4 - c^2 a^2} = \frac{z}{c^4 - a^2 b^2} = K$$

$$\Rightarrow \quad \left. \begin{array}{l} x = K(a^4 - b^2 c^2) \\ y = K(b^4 - c^2 a^2) \\ z = K(c^4 - a^2 b^2) \end{array} \right\} I$$

Equation (3) is $z^2 - xy = c^2$. Substituting the values of x, y, z. we get.

$$K^2 (c^4 - a^2 b^2)^2 - K^2 (a^4 - b^2 c^2)(b^4 - c^2 a^2) = c^2$$

$$K^2 [(c^4 - a^2 b^2)^2 - (a^4 - b^2 c^2)(b^4 - c^2 a^2)] = c^2$$

$$K^2 [c^8 + a^4 b^4 - 2a^2 b^2 c^4 - (a^4 b^4 - c^2 b^6 - b^6 c^2 + c^4 a^2 b^2)] = c^2$$

$$K^2 [c^8 - 3a^2 b^2 c^4 + c^2 a^6 + b^6 c^2] = c^2$$

$$\Rightarrow K^2 [a^6 + b^6 + c^6 - 3a^2 b^2 c^2] = 1$$

$$K^2 = \frac{1}{a^6 + b^6 + c^6 - 3a^2b^2c^2}$$

$$K = \frac{\pm 1}{\sqrt{a^6 + b^6 + c^6 - 3a^2b^2c^2}}$$

Substituting this value of K in I we get the solutions for the equations.

8. $x + y + z = 9$ is the first equation squaring both sides we get

$$(x + y + z)^2 = 81$$

$$x^2 + y^2 + z^2 + 2(xy + yz + zx) = 81$$

Using the second equation we get

$$xy + yz + zx = 26 \qquad\qquad\qquad \ldots(1)$$

We know that

$$x^3 + y^3 + z^3 - 3xyz = (x + y + z)(x^2 + y^2 + z^2 - xy - yz - zx)$$

Using the three equations and (1) we get

$$99 - 3xyz = 9(29 - 26)$$

$$\Rightarrow \qquad xyz = 24 \qquad\qquad\qquad \ldots(2)$$

$$(1) \qquad \Rightarrow \qquad xy + yz + zx = 26$$

$$x(y + z) + yz = 26$$

Using the first equation and (2) we get

$$x(9 - x) + \frac{24}{x} = 26$$

$$\Rightarrow \qquad x^3 - 9x^2 + 26x - 24 = 0$$

$$\Rightarrow \qquad (x - 2)(x - 3)(x - 4) = 0$$

$$x = 2, 3, 4$$

From the given equations we get

$x = 2, \quad y = 3, \quad z = 4$

$x = 2, \quad y = 4, \quad z = 3$

$$x = 3, \quad y = 2, \quad z = 4$$

$$x = 3, \quad y = 4, \quad z = 2$$

$$x = 4, \quad y = 2, \quad z = 3$$

$$x = 4, \quad y = 3, \quad z = 2$$

9.   Add two equations we get

$$x^2 + y^2 + y(x+1) + x(y+1) = 30$$

$$x^2 + y^2 + xy + y + xy + x = 30$$

$$(x^2 + y^2 + 2xy) + (x + y) - 30 = 0$$

$$(x + y)^2 + (x + y) - 30 = 0$$

$$(x + y - 5)(x + y + 6) = 0$$

$$x + y = 5 \quad or \quad x + y = -6 \qquad \ldots (2)$$

Take the given equations.

Subtract the second from the first.

$$x^2 + y(x+1) - y^2 - x(y+1) = 4$$

$$x^2 - y^2 + xy + y - xy - x = 4$$

$$x^2 - y^2 - (x - y) = 4$$

$$(x + y)(x - y) - (x - y) = 4$$

$$\Rightarrow \qquad (x - y)[x + y - 1] = 4 \qquad \ldots (i)$$

Case (I): from (i) we have $x + y = 5$

(i) $\qquad \Rightarrow \qquad (x - y)[5 - 1] = 4$

$\qquad\qquad \Rightarrow \qquad x - y = 1$

and $\qquad x + y = 5 \Rightarrow x = 3, \ y = 2$

Case (II): From (2) $x + y = -6$

(I) $\qquad \Rightarrow \qquad (x - y)[-6 - 1] = 4$

$$\Rightarrow \quad x - y = \frac{4}{7}$$

$$x + y = -6 \Rightarrow x = -\frac{23}{7}, \ y = -\frac{19}{7}$$

5. POLYNOMIALS

EXERCISE - 2
SOLUTIONS

1. Let the roots of the equation

$$x^4 + px^3 + qx^2 + rx + s = 0,$$

be $\alpha, \beta, \gamma, \delta$, so that $\alpha > 0, \beta > 0, \gamma > 0, \delta > 0$.

Now, $\qquad \Sigma\alpha = -p$,

$$\Sigma\alpha\beta = q,$$

$$\Sigma\alpha\beta\gamma = -r,$$

$$\alpha\beta\gamma\delta = s$$

(i) $pr = \Sigma\sigma\Sigma\alpha\beta\gamma$

By the inequality of the means

$$\frac{1}{4}\Sigma\alpha \geq (\alpha\beta\gamma\delta)^{\frac{1}{4}}, \qquad\qquad \dots (1)$$

$$\frac{1}{4}\Sigma\alpha\beta\gamma \geq (\alpha\beta\gamma\delta)^{\frac{3}{4}} \qquad\qquad \dots (2)$$

By multiplying corresponding sides of the above inequalities, we have

$$\frac{1}{16}\Sigma\alpha\Sigma\alpha\beta\gamma \geq \alpha\beta\gamma\delta.$$

i.e., $\qquad pr - 16s \geq 0$

Equality holds in (3) $\Leftrightarrow$ equalities hold in both (1) and (2) $\Leftrightarrow$ $\alpha, \beta, \gamma, \delta$ are all equal, and $\alpha\beta\gamma, \alpha\beta\delta, \alpha\gamma\delta, \beta\gamma\delta$ are all equal $\Leftrightarrow \alpha = \beta = \gamma = \delta$.

(ii) $q^2 = (\Sigma\alpha\beta)^2 = \left[6(\alpha\beta.\alpha\gamma.\alpha\delta.\beta\gamma.\beta\delta.\gamma\delta)^{\frac{1}{6}}\right]^2$

$$= 36\,\alpha\beta\gamma\delta$$

$$= 36$$

226

Therefore, $\qquad q^2 \geq 36s$

Equality holds if and only if $\alpha\beta$, $\alpha\gamma$, $\alpha\delta$, $\beta\gamma$, $\beta\delta$, $\gamma\delta$ are all equal. i.e., if and only if $\alpha = \beta = \gamma = \delta$.

2. Suppose that we can express $f(x)$ as a product $f(x) = p(x)q(x)$, where $p(x)$ and $q(x)$ are both polynomials with integral co-efficients and degree less than 4. Two different cases arise:

Case I. One of the polynomials, say $p(x)$, is of degree one and the other, say q(x), is of degree 3.

Let $\qquad f(x) = (x+a)(x^2 + bx^2 + cx + d) \qquad$ ... (A)

$$a + b = 26 \qquad\qquad \text{... (i)}$$
$$ab + c = 52 \qquad\qquad \text{... (ii)}$$
$$ac + d = 78 \qquad\qquad \text{... (iii)}$$
$$ad = 1989 \qquad\qquad \text{... (iv)}$$

From (iv), we have $ad = 9 \times 13 \times 17$, so that exactly one of the numbers a, d is divisible by 13. If $13|a$ $13|a$ and $13 + d$, we have a contradiction from (iii) since RHS of (iii) is a multiples of 13. One the other hand if $13 + a$ and $13|a$, then (iii) $\Rightarrow 13|c$, which in turn implies that $13|ab$. Since $13 + a$, therefore $13|b$. From (i) it follows that $13|a$, which is a contradiction.

Thus, we find that $f(x)$ cannot be expressed in form (A).

Case II. Let $\qquad f(x) = (x^2 + ax + b)(x^2 + cx + d) \qquad$ ... (B)

Comparing co-efficients of like powers of x on both sides, we have

$$a + c = 26 \qquad\qquad \text{... (v)}$$
$$ac + b + d = 52 \qquad\qquad \text{... (vi)}$$
$$ad + bc = 78 \qquad\qquad \text{... (vii)}$$
$$bd = 1989 \qquad\qquad \text{... (viii)}$$

As before, from (viii) we find that exactly one of b, d is a multiple of 13.

227

If $13\,|\,b$, $13+d$, from (vii) it follows that $13\,|\,a$. Since $13\,|\,a$, $13\,|\,b$, but $13+d$, contradiction follows from (vi).

If $13+d$, $13\,|\,d$, from (vii) it follows that $13|c$. Since $13\,|\,c$, $13\,|\,d$, but $13+b$, contradiction again follows from (vi).

Thus we find that $f(x)$ cannot be expressed in form (B).

The proof is now complete.

3. Let us suppose that all the six roots of the given equation are real.

Let us denote the roots by α, β, γ, δ, σ and μ.

Now, $\Sigma\alpha = 0$, $\Sigma\alpha\beta = 0$,

$\therefore$ $\Sigma\alpha^2 = (\Sigma\alpha)^2 - 2\Sigma\alpha\beta = 0$

Since $\alpha^2 + \beta^2 + \gamma^2 + \delta^2 + \lambda^2 + \mu^2 = 0$, and α, β,..... are all real, therefore $\alpha = \beta = \gamma = \lambda = \mu = 0$. Consequently we must have $a = b = c = d = 0$, which is impossible since we are given that a, b, c, d are not all zero.

Hence the roots of the given equation cannot be all real.

4. Suppose that there exist polynomials $f(x)$, $g(x)$ with integer co-efficients such that

$$f(x)g(x) = (x - a_1)^2(x - a_2)^2 \,...... \, (x - a_n)^2 + 1 \qquad ...\,(1)$$

Since RHS is always positive, therefore $f(x)$ can never vanish and so its sign never change. Similarly $g(x)$ can never vanish and its sign never changes. Since $f(x)$, $g(x)$ is always positive, we can assume without loss of generality, that $f(x)$ and $g(x)$ both are always positive.

Substituting $x = a_1, a_2, ...,$ in (1), we find that

$$f(a_1)g(a_1) = 1, \; f(a_2)g(a_2) = 1, \; ...., \; f(a_n)g(a_n) = 1$$

Since $f(a_1).....$, $f(a_n)$ are all positive integers, it follows that

$$f(a_1) = f(a_2) = ...= f(a_n) = 1$$

Similarly, $g(a_1) = g(a_2) = ...= g(a_n) = 1$

Since $f(x) - 1, g(x) - 1$ vanish when $x = a_1, a_2, \ldots\ldots, a_n$, therefore by factor theorem

$$f(x) - 1 = p(x)(x - a_1)(x - a_2)\ldots\ldots(x - a_n),$$

$$g(x) - 1 = q(x)(x - a_1)(x - a_2)\ldots\ldots(x - a_n),$$

where $p(x), q(x)$ are polynomials with integer co-efficients.

Since $f(x)g(x)$ is of degree $2n$, $p(x)$ and $q(x)$ must be both constants. Suppose $p(x) = a, q(x) = b$.

Then, $\quad f(x) = a(x - a_1)(x - a_2)\ldots\ldots(x - a_n) + 1$

$$g(x) = b(x - a_1)(x - a_2)\ldots\ldots(x - a_n) + 1$$

Substitutting the expressions for $f(x)$ and $g(x)$ in (1), we find that

$$ab = 1, a + b = 0.$$

Since there do not exist any real numbers, a, b satisfying there conditions (these conditions imply $a^2 = -1, b^2 = -1$), therefore there is a contradiction, and the given polynomial cannot be expressed as the product of two polynomials with integer co-efficients.

5. Suppose it is possible that $P(a) = b, P(b) = c, P(c) = a$. Since $P(x) - P(b)$ vanishes when $x = b$, therefore by the factor theorem $x - b$ must be a factor of $P(x) - P(b)$. In particular, $a - b$ must divide $P(a) - P(b)$, i.e., $b - c$. Similarly $b - c$ must divide $a - b$. Therefore $b - c = \pm(a - b)$. Since a, b, c are all distinct, therefore we cannot have $b - c = -(a - b)$. Since a, b, c are all distinct, therefore we cannot have $b - c = -(a - b)$. Consequently $b - c = a - b$. This means that $b = \dfrac{1}{2}(a + c)$, so that b must be between a and c. Similarly we can conclude that c lies between a and b. It is obvious that both the conclusions cannot be simultaneously true. Hence our supposition must be wrong, and it cannot be possible that $P(a) = b, P(b) = c, P(c) = a$.

6. Suppose there exists an integer b such that $f(b) = 9$.

Since $f(x) - f(b) = 0$, when $x = b$, therefore $x - b$ must divide $f(x) - f(b)$. In particular, $a_1 - b$ must divide $f(a) - f(b)$. We are given that $f(a_1) - f(b) = -7$. Therefore $(a_1 - b) | (-7)$. Since the only divisors of -7 and $-1, -1, +1, +7$, therefore $a_1 - b$ must have one of the these values.

Similarly $a_2 - b$, $a_3 - b$, $a_4 - b$, $a_5 - b$ must take one of the values $\pm 1, \pm 7$. By the pigeon hole principle, (at least) some two of the five numbers $a_i - b, i = 1, 2, ..., 5$ must be equal. This is not possible since the a_i's are all distinct. Consequently $f(x)$ cannot be equal to 9 for any integer b.

7. Let the zeros of $p(b)$ be α, β, so that

$$p(x) = (x - \alpha)(x - \beta)$$

Then, $p(n) = (n - \alpha)(n - \beta)$, $p(n+1) = (n+1-\alpha)(n+1-\beta)$. We have to show that $p(n)p(n+1)$ can be written as $(t-\alpha)(t-\beta)$ for some integer t (which will depend upon n).

$$p(n)p(n+1) = (n-\alpha)(n-\beta)(n+1-\alpha)(n+1-\beta)$$

$$= \left\{(n-\alpha)(n+1-\beta)\right\} \left\{(n-\beta)(n+1-\alpha)\right\}$$

$$= \left\{n(n+1) - (n+1-\beta) - \alpha + \alpha + \alpha\beta\right\} \times$$

$$\left\{n(n+1) - n(\alpha+\beta) - \beta + \alpha\beta\right\},$$

$$= \left\{n(n+1) + na + b - \alpha\right\}\left\{n(n+1) + na + b - \beta\right\}$$

$$= (t-\alpha)(t-\beta), \text{ where } t = n(n+1) + an + b$$

$$= p(t).$$

Thus, $p(n)p(n+1)$ can be written as $p(M)$ for $M = n(n+1) + an + b$.

8. Let the quotient be $q(x)$ and remainder be $r(x)$ when $f(x)$ is

divided by $(x-3)(x-1)^2$. Since the divisor is a polynomial of degree 3, the remainder must be a polynomial of degree at most 2, i.e., it must be of the form $ax^2+bx+c,$ where a,b,c are some rational numbers.

We can write $ax^2+bx+c = a[(x-1)+1]^2+b[(x-1)+1]+c,$

$$= (x-1)^2+(2a+b)(x-1)+a+b+c.$$

By division algorithm,

$$f(x)=q(x)\,(x-3)(x-1)^2+a(x-1)^2+(2a+b)(x-1)+a+b+c$$
$$\qquad\qquad\qquad\qquad\qquad\qquad\qquad\qquad \text{...(1)}$$

since $f(x)$ leaves a remainder 15 when divided by $x-3$, therefore by remainder theorem $f(3)=15$. Putting $x=3$ in (1) throughout and using the fact that $f(3)=15$, we have

$$15=9a+3b+c. \qquad\qquad\qquad \text{... (2)}$$

also from (1), we find that the remainder when $f(x)$ is divided by $(x-1)^2$ is $(2a+b)(x-1)+(a+b+c)$. Since this is given to be $2x+1$, we have $(2a+b)(x-1)(a+b+c)=2x+1$.

Putting $x=1$, throughout, we have

$$a+b+c=3 \qquad\qquad\qquad\qquad \text{... (3)}$$

Also putting $x=0$ throughout, we have

$$-a+c=1 \qquad\qquad\qquad\qquad \text{... (4)}$$

From (2), (3) and (4), we have $a=2,\ b=-2,\ c=3$. Therefore the remainder $= ax^2+bx+c = 2x^2-2x+3$.

9.   Since $y-1$ divides y^n-1 for every positive integer n, therefore by writing $x^{pq}-1$, we find that x^p-1 divides $x^{pq}-1$. Similarly, x^q-1 also divides $x^{pq}-1$.

Since p and q are prime to each other, therefore the GCD of x^p-1 and x^q-1 is $x-1$. Consequently $x^p-1=(x-1)f(x)$,

$x^q - 1 = (x-1)g(x)$, where $f(x), g(x)$ have no common factor. Now $f(x), g(x), (x-1)$ have no factor in common, and each of them divides $x^{pq} - 1$, therefore $(x-1)(f(x)g(x)$ divides $x^{pq} - 1$. Consequently $(x-1)^2 f(x)g(x)$ divides $(x^{pq} - 1)(x-1)$, i.e., $(x^p - 1)(x^q - 1)$ divides $(x^{pq} - 1)(x-1)$.

6. FUNCTIONAL EQUATIONS

EXERCISE - 2
SOLUTIONS

1. Consider the conditions (a) and (c)

   (i.e.) $f(f(n)) = n \ \forall \ n \in Z$ and $f(0) = 1$

   In (a), put $n = 0 \Rightarrow f(f(0)) = 0$

   But $f(0) = 1 \Rightarrow f(1) = 0$; (b) is $f(f(n+1) + 2) = n$.

   Put $n = -1$ we get $f(f(1) + 2) = -1 \Rightarrow f(2) = -1$ ... (1)

   $(a) \Rightarrow f(f(2) = 2 \Rightarrow f(-1) = 2$ ... (2)

   In (b), but $n = -2 \ \ f(f(0) + 2) = -2$.

   Since $f(0) = 1$, we have $f(2) = -2$ ... (3)

   $f(f(3)) = 3 \Rightarrow f(-2) = 3$

   The above pattern suggests that $f(n) = -n + 1$ ... (4)

   Let us prove (4) by induction on n.

2. We have by conditions

$$f(x, x+y) = \left(\frac{x+y}{y} \right) f(x, y)$$

Replace $x + y$ by z, we get

$$f(x, z) = \frac{z}{z - x} f(x, z - x)$$

which is valid when $z - x > 0$

$$f(52, 14) = f(14, 52) = \frac{52}{38} f(14, 38)$$

$$= \frac{52}{38} \frac{38}{24} f(14, 24)$$

$$= \frac{52}{24} \cdot \frac{24}{10} f(14, 10)$$

233

$$= \frac{26}{5} f(10, 14) = \frac{26}{5} \frac{14}{4} f(10, 4)$$

$$= \frac{91}{5} f(4, 10) = \frac{91}{5} \cdot \frac{10}{6} f(4,6)$$

$$= \frac{91}{3} \frac{6}{2} f(4, 2) = 91 f(2, 4)$$

$$= 91 \frac{4}{2} f(2,2) = 364 .$$

3.
$$g(x) = \frac{ax+b}{7x-b}$$

$$g(g(x)) = \frac{ag(x)+6}{7g(x)-6} = x$$

To find a, b,

$$ag(x) + b = a \left[\frac{ax+b}{7x-6} \right] + b = \frac{(a^2 + 7b)x + ab - 6b}{7x-6}$$

$$7g(x) - 6 = 7 \left(\frac{ax+b}{7x-6} \right) - 6 = \frac{(7a-42)x + 7b + 36}{7x-6}$$

When $x \neq \dfrac{6}{7}$ and $7g(x) \neq 6$ we should have

$$x = \frac{ag(x)+6}{7g(x)-6} = \frac{(a^2 + 7b)x + ab - 6b}{(7a-42)x + 7b + 36}$$

(i.e.,) $(7a - 42)x^2 + (7b + 36 - a^2 - 7b)x + 6b - ab = 0$

This has to satisfied by more than 2 values of x

$\Rightarrow \qquad 7a - 42 = 36 - a^2 = 6b - ab = 0$

If $7b + 36 = 0$ then $7g(x) = 6$ for any x.

Thus $a = 6$ and b is any real number other than $-\dfrac{36}{7}$

4. (a) As 1996 and 1999 are of the forms $4m$ and $4m + 3, m \geq 1$ we will try to prove the more general problem $f(4m + 3) > f(4m)$ where $m \geq 1$. Now consider any triangle with sides x, y, z and perimeter $N = 4m$. Adding 1 to the lengths x, y, z we get the lengths $x + 1, y + 1, z + 1$ which are the sides of the triangles with perimeter $4m + 3$ for $x + y + z = 4m$. But we have also triangles with sides $(1, 2m + 1, 2m + 1)$, $(2, 2m, 2m + 1), \ldots\ldots, (m + 1, m + 1, 2m + 1)$ and the perimeter of these triangles are $4m + 3$. Also subtracting 1 from the sides of these triangles we do not get any triangle corresponding to these triplets (x, y, z) for in all these cases $x + y = z$. Hence $f(4m + 3) > f(4m)$. By substituting $m = 499$, we get $f(1999) > f(1996)$.

(b) As $2000 = 4 \times 500$ and $1997 = 4 \times 500 - 3$, we will prove the more general result $f(4m) = f(4m - 3)$ for any natural number $m \geq 2$. Let S and T be the sets of triangles with perimeters $4m - 3$ and $4m$ respectively. Let x, y, z be the sides of a triangle with $x \leq y \leq z$ and $x + y > z$. Let the triplet (x, y, z) denote such a triangle with sides x, y, z.

We have $(x, y, z) \in S \Rightarrow (x + y > z)$ where $x + y + z = 4m - 3$.

$\Rightarrow \qquad (x + 1) + (y + 1) > z + 1$

$\Rightarrow \qquad (x + 1, y + 1, z + 1) \in T \qquad\qquad \ldots (A)$

as $(x + 1) + (y + 1) + (z + 1) = x + y + z + 3 = (4m - 3) + 3 = 4m$.

Also, if $(x, y, z) \in T$, then $(x + y) + z = 4m$ (an even numbers) $\Rightarrow x + y$ and z are of the same parity (i.e. either both even or both odd).

$\Rightarrow$ min. value of $(x + y) - z = 2$

$\Rightarrow \qquad (x + y) \geq z + 2$

$\Rightarrow \qquad (x - 1) + (y - 1) \geq (z - 1) + 1$

$\Rightarrow \qquad (x - 1) + (y - 1) > z - 1$ for all $(x, y, z) \in T$

$$\Rightarrow \quad (x-1, y-1, z-1) \in S \ \forall \ (x, y, z) \in T \qquad \text{... (B)}$$

So by A and B we have established a bijection between S and T.

i.e., $f(4m-3) = f(4m) \Rightarrow f(1997) = f(2000)$ for $m = 500$.

5. Given: $f(2+x) = f(2-x)$ and $f(7+x) = f(7-x)$

$$\begin{aligned} \text{Now} \quad f(2+x) &= f(2-x) \\ &= f(7-(5+x)) \\ &= f(7+5+x) \\ &= f(12+x) \end{aligned}$$

Here x is any real number. Replace x by $x-2$ we get

$$f(x) = f(x+10) \ \forall \ x \in R \qquad \text{... (1)}$$

Given $f(0) = 0$

$$\Rightarrow 0 = f(2-2) = f(2+2) = f(4) \qquad \text{... (2)}$$

$$(1) \quad \Rightarrow \quad f(0) \quad = \quad f(10) = 0$$

$$(1) \quad \Rightarrow \quad f(10) \quad = \quad f(20) = 0$$

$$\Rightarrow \quad f(10) \quad = \quad f(20) = f(30) = \dots = f(10n) = 0$$

$$(1) \quad \Rightarrow \quad f(4) \quad = \quad f(10+4) = 0 \quad \text{(by (2))}$$

$$\Rightarrow \quad f(4) \quad = \quad f(14) = f(14+10)$$

$$= \quad f(20+4) = \dots f(10n+4) = 0$$

$$\therefore \quad f(10n) \quad = \quad f(10n+4) = 0 \ \forall \ n \in I$$

In the interval $[-2002, 2002]$ we have 401 numbers of the form $10n$ and 400 numbers of the term $10n+4$.

So there are 801 numbers in $[-2002, 2002]$ satisfy $f(x) = 0$.

$$f(x) = \begin{cases} 0 & when \ x = 10n \\ & or \ x = 10n+4 \\ \neq 0 & otherwise \end{cases}$$

We have exactly 801 numbers x such that $f(x) = 0$ in the interval

236

$-2002 \le x \le 2002$.

6. Let $g(x) = xf(x) - 1$. Then $g(k) = kf(k) - 1 = 0$ for $k = 1, 2, ... 99$ because they are the roots of $g(x)$.

Let $g(x) = a_0(x-1)(x-2) .... (x-99)$

$\therefore$ $g(100) = a_0(100-1)(100-2).....(100-99)$

$= a_0 99.98 ......1 = a_0(99!)$

Now we have to evaluate the constant a_0.

$$g(0) = a_0(-1)(-2) ....(-99)$$

$$g(0) = a_0(-1)^{99} 99! = -a_0 99!$$

$\therefore$ $g(0) = -a_0 99!$

But $g(0) = 0 f(0) - 1 = -1 \Rightarrow -a_0 99! = -1 \Rightarrow a_0 = \dfrac{1}{99!}$

$(1) \Rightarrow g(100) = 1$

But $g(100) = 100 f(100) - 1 = 1 \Rightarrow 100 f(100) = 2$

$f(100) = \dfrac{2}{100} = \dfrac{1}{50}$

8. Let $x + y = h, \ x^2 + y^2 = k$

$$(x+y)^2 = x^2 + y^2 + 2xy \Rightarrow xy = \frac{h^2 - k}{2}$$

We have $x + y = h$ and $xy = \dfrac{h^2 - k}{2} = 0$

$\Rightarrow x, y$ are the roots of the equation

$$t^2 - ht + \frac{h^2 - k}{2} = 0.$$

This quadratic has positive roots if $D \ge 0$.

(i.e.,) $h^2 - 2(h^2 - k) \ge 0$

Since x, y are positive reals h is also.

The roots are $h \pm \sqrt{h^2 - 2(h^2 - k)}$.

Since the roots are positive.

$$h > \sqrt{h^2 - 2(h^2 - k)} \qquad or \qquad h^2 > h^2 - 2(h^2 - k)$$

Thus the quadratic equation has two positive roots if

(1) $\qquad h > 0$

(2) $\qquad h^2 - 2(h^2 - k) \geq 0 \Rightarrow k \geq \dfrac{h^2}{2}$

(3) $\qquad h^2 > h^2 - 2(h^2 - k) \Rightarrow k < h^2$

(i.e.,) $\qquad \dfrac{h^2}{2} \leq k < h^2$ $\qquad\qquad$ ... (4)

We have $f(h) = f(k)$

$\Rightarrow f$ must be constant in the interval $\left[\dfrac{h^2}{2}, h^2 \right]$

Then $\quad$ .... $A_{-4}, A_{-3}, A_{-2}, A_{-1}, A_0, A_1, A_2, A_3, A_4, \ldots\ldots$

cover the set of all positive numbers.

Further we observe that A_n and A_{n+1} overlap.

Hence f must be a constant function on the set of positive real numbers.

8. $\quad f_1(11) \quad = \quad (1+1)^2 \quad = \quad 4$

$\quad f_2(11) \quad = \quad f_1(4) \quad = \quad 16$

$\quad f_3(11) \quad = \quad f_1(16) \quad = \quad (1+6)^2 = 49$

$\quad f_4(11) \quad = \quad f_1(49) \quad = \quad (4+9)^2 = 169$

$\quad f_5(11) \quad = \quad f_1(169) = \quad (1+6+9)^2 = 256$

$\quad f_6(11) \quad = \quad f_1(256) = \quad (2+5+6)^2 = 169$

$f_7(11) = f_1(169) = 256$

$f_8(11) = f_1(256) = 169$

$\therefore$ For $n \geq 6$, $f_n(11) = 169$ and $f_{1998}(11) = 169$

7. INEQUALITIES

EXERCISE - 2
SOLUTIONS

1. The AM of x, y, z is $\dfrac{x+y+z}{3}$

    The Gm of x, y, z is $\sqrt[3]{xyz}$

    The AM of $\dfrac{1}{x}, \dfrac{1}{y}, \dfrac{1}{z}$ is $\dfrac{\frac{1}{x}+\frac{1}{y}+\frac{1}{z}}{3}$

    The Gm of $\dfrac{1}{x}, \dfrac{1}{y}, \dfrac{1}{z}$ is $\sqrt[3]{\dfrac{1}{xyz}}$

    Applying $AM \geq GM$ for the first and second terms we have

    $$\dfrac{x+y+z}{3} \geq \sqrt[3]{xyz}$$

    $$\dfrac{\frac{1}{x}+\frac{1}{y}+\frac{1}{z}}{3} \geq \sqrt[3]{\dfrac{1}{xyz}}$$

    Multiplying we get,

    $$\dfrac{(x+y+z)\left(\frac{1}{x}+\frac{1}{y}+\frac{1}{z}\right)}{9} \geq \sqrt[3]{xyz\dfrac{1}{xyz}}$$

    $$\Rightarrow \quad (x+y+z)\left(\dfrac{1}{x}+\dfrac{1}{y}+\dfrac{1}{z}\right) \geq 9$$

2. We have

    $$(a+b+c)^2 = a^2+b^2+c^2+2(ab+bc+ca),$$

    Since $a+b+c=1$

    We have $1 = a^2+b^2+c^2+2(ab+bc+ca)$

    $$\Rightarrow \quad a^2+b^2+c^2 = 1-2(ab+bc+ca) \qquad \ldots (1)$$

    We have $(a-b)^2+(b-c)^2+(c-a)^2 \geq 0$

$\Rightarrow \qquad a^2 + b^2 + c^2 \geq ab + bc + ca$

or $\qquad (a+b+c)^2 - 2(ab+bc+ca) \geq ab+bc+ca$

$\Rightarrow \qquad 1 - 2(ab+bc+ca) \geq ab+bc+ca$

$\Rightarrow \qquad 1 \geq 3(ab+bc+ca)$

$\Rightarrow \qquad 2 \geq 6(ab+bc+ca)$

$\Rightarrow \qquad 1 - 2(ab+bc+ca) \geq 4(ab+bc+ca) - 1$

$\Rightarrow \qquad a^2 + b^2 + c^2 \geq 4(ab+bc+ca) - 1$

3. Since a, b, c are the sides of a triangle we have $a, b, c > 0$ adn we know that the sum of any two sides of a triangle is greater than the third side

$\Rightarrow \qquad a+b-c, \ b+c-a, \ c+a-b$ are all positive.

Applying $AM \geq GM$ to the quantities,

$$\frac{a}{c+a-b}, \frac{b}{a+b-c}, \frac{c}{b+c-a}$$

we have $\qquad \dfrac{a}{c+a-b} + \dfrac{b}{a+b-c} + \dfrac{c}{b+c-a}$

$$\geq 3 \sqrt[3]{\frac{abc}{(c+a-b)(a+b-c)(b+c-a)}}$$

we have $\qquad a^2 \geq a^2 - (b-c)^2$

or $\qquad a^2 \geq (a+b-c)(a-b+c)$

$\parallel^{ly} \qquad b^2 \geq (b+c-a)(b-c+a)$

$c^2 \qquad \geq \qquad (c+a-b)(c-a+b)$.

Multiplying together we get

$a^2 b^2 c^2 \geq (a+b-c)^2 (b+c-a)^2 (c+a-b)^2$

Taking the positive root both sides we have

$\qquad abc \geq (a+b-c)(b+c-a)(c+a-b)$

241

$$\Rightarrow \quad \frac{abc}{(a+b-c)(b+c-a)(c+a-b)} \geq 1$$

$$\Rightarrow \quad \sqrt[3]{\frac{abc}{(a+b-c)(b+c-a)(c+a-b)}} \geq 1$$

$$\Rightarrow \quad \frac{a}{c+a-b}+\frac{b}{a+b-c}+\frac{c}{b+c-a} \geq 3$$

4.   For positive reals x, y applying $AM \geq GM$ $\dfrac{x+y}{2} \geq \sqrt{xy}$

Since, $x+y=1$

we have, $\dfrac{1}{2} \geq \sqrt{xy}$

$$\Rightarrow \quad 1 \geq 2\sqrt{xy}$$

$$1 \geq 4xy$$

$$2 \geq 8xy$$

$$1+1 \geq 8xy$$

(i.e.,) $(x+y+1) \geq 8xy$

$$x+y+xy+1 \geq 9xy$$

$$(x+1)(y+1) \geq 9xy$$

$$\frac{(x+1)(y+1)}{xy} \geq 9$$

$$\Rightarrow \quad \left(\frac{x+1}{x}\right)\left(\frac{y+1}{y}\right) \geq 9$$

$$\Rightarrow \quad \left(1+\frac{1}{x}\right)\left(1+\frac{1}{y}\right) \geq 9$$

5.   To prove that $a^2+b^2+c^2+\sqrt{12abc} \leq 1$

To prove that $\sqrt{12abc} \le 1 - (a^2 + b^2 + c^2)$ ... (1)

At this stage, let us use $a + b + c = 1$

We have to prove $\sqrt{12abc} \le (a+b+c)^2 - (a^2+b^2+c^2)$.

(i.e.,) $\sqrt{12abc} \le 2(ab + bc + ca)$

To prove that $12abc \le 4(ab + bc + ca)^2$

To prove that $3abc \le (ab + bc + ca)^2$ ... (2)

Now, $(ab - bc)^2 + (bc - ca)^2 + (ca - ab)^2 \ge 0$

$\Rightarrow$ $2(a^2b^2 + b^2c^2 + c^2a^2) - 2abc(a + b + c) \ge 0$

$\Rightarrow$ $a^2b^2 + b^2c^2 + c^2a^2 \ge abc$

$\Rightarrow$ $(ab + bc + ca)^2 - 2abc(a + b + c) \ge abc$

$\Rightarrow$ $(ab + bc + ca)^2 \ge 3abc$

which is precisely (2).

6.   Consider the positive reals ab, ac, ad, bc, bd, cd.

Apply $AM - GM$ inequality

$$\frac{ab + ac + ad + bc + cd + cd}{6} \ge \sqrt[6]{ab.ac.ad.bc.bd.cd}$$

$\Rightarrow$ $ab + ac + ad + bc + bd + cd \ge 6\sqrt{abcd}$

(2) $\Rightarrow$ $abcd \ge 27 + 6\sqrt{abcd}$

$\Rightarrow$ $abcd - 6\sqrt{abcd} - 27 \ge 0$

$\Rightarrow$ $\left(\sqrt{abcd} - 9\right)\left(\sqrt{abcd} + 3\right) \ge$

$\Rightarrow \sqrt{abcd} \ge 9$ ... (1)

Now, $\dfrac{a + b + c + d}{4} \ge \sqrt[4]{abcd}$

But, $a + b + c + d = 12$ by equation (1).

$$3 \geq \sqrt[4]{abcd}$$

$$\Rightarrow \qquad 9 \geq \sqrt{abcd} \qquad\qquad\qquad ... (2)$$

From 1 and 2 we get $\sqrt{abcd} = 9$ or $abcd = 81$. Thus we have, $AM = GM$.

This can happen when $\qquad\qquad a = b = c = d$

$$\Rightarrow \qquad a = b = c = d = 3$$

9. PROGRESSIONS

EXERCISE - 2
SOLUTIONS

1. Clearly the given series is an A.P with

   $a = 1017, d = 1035 - 1017 = 18$ and $l = 9999$

   Let it contain n terms

   $\therefore \qquad 9999 = T_n = a + (n-1)d$

   $\Rightarrow \qquad 8982 = (n-1) \times 18 \Rightarrow n = 500$

   $\therefore \qquad S_n = \dfrac{n}{2}(a+l) = \dfrac{500}{2}(1017 + 9999)$

   $= 250 \times 11,016 = 27,54,000$

2. $a_2 - a_6 + a_4 = -7 \Rightarrow a + d - (a+5d) + (a+3d) = -7$

   $a - d = -7 \qquad \qquad \qquad \qquad \text{... (1)}$

   $a_8 - a_7 = 2a_4 \Rightarrow a + 7d - (a+6d) = 2(a+3d)$

   $\Rightarrow d = 2a + 6d \Rightarrow 2a + 5d = 0 \qquad \text{... (2)}$

   Solving (1) and (2) we get $a = -5, d = 2$

   $\therefore \qquad$ Progression = $\{-5, 3, 1, 3, 5, \ldots\ldots\ldots$

3. The number of two digits which are divisible by 7 are
   14, 21, 28 ... 98

   This is an A.P. with a = 14 and d = common difference
   (C.D.) = 21 − 14 = 7

   Let this A.P. contain n terms. By hypothesis

   $98 = T_n = a + (n-1)d = 14 + (n-1)7$

   $\Rightarrow 84 = (n-1).7 \Rightarrow n = 13$

   Thus the number of two digits divisible by 7 is 13

4. 1. $t_n = a(r)^{n-1} = 1.\left(\dfrac{1}{3}\right)^{n-1} = \left(\dfrac{1}{3}\right)^{n-1}$

2. $S_n = \dfrac{a(1-r^n)}{1-r}$ $\left(\because r = \dfrac{1}{3} < 1 \right)$

$$= \dfrac{1.\left(1-(1/3)^n\right)}{1-1/3} = \dfrac{3}{2}\left(1-\dfrac{1}{3^n}\right)$$

5.   Here a = 243, and $r = \dfrac{324}{243} = \dfrac{4}{3} > 1$

$\therefore \qquad S_n = \dfrac{a(r^n - 1)}{r - 1} = \dfrac{243\left[\left(\dfrac{4}{3}\right)^n - 1\right]}{\dfrac{4}{3} - 1}$

$$= 35\left[\dfrac{4^n}{3^n} - 1\right]\bigg/\left(\dfrac{1}{3}\right) = 3^6 \dfrac{4^n - 3^n}{3^n} = 3^{6-n}.(4^n - 3^n)$$

6.   Let $S = \displaystyle\sum_{n=1}^{7}(2^n + 3^{n-1}) = \sum_{n=2}^{7} 2^n + \sum_{n=2}^{7} 3^{n-1}$

$$= (2^2 + 2^3 + \ldots + 2^7) + (3^1 + 3^2 + \ldots + 3^6)$$

$$= \dfrac{2^2(2^6 - 1)}{2 - 1} + \dfrac{3^1(3^6 - 1)}{3 - 1}$$

$$= 4(64 - 1) + \dfrac{3}{2}(729 - 1) = 1344$$

7.   Let n be the number of trms required

Here, $a = \sqrt{3}$ and $r = \sqrt{3}$

Also, $S_n = 39 + 13\sqrt{3}$

Using the formula $S_n = \dfrac{a(r^n - 1)}{r - 1}$, we get

$$39 + 13\sqrt{3} = \frac{\sqrt{3}\left[\left(\sqrt{3}\right)^n - 1\right]}{\sqrt{3} - 1}$$

$$\Rightarrow 13\sqrt{3}\left(\sqrt{3} + 1\right)\left(\sqrt{3} - 1\right) = \left(\sqrt{3}\right)^{n+1} - \sqrt{3}$$

$$\Rightarrow 26\sqrt{3} = \left(\sqrt{3}\right)^{n+1} - \sqrt{3}$$

$$\Rightarrow \left(\sqrt{3}\right)^{n+1} - 27\sqrt{3}$$

$$\Rightarrow (n+1)/2 = 7/2 \Rightarrow n = 6$$

8. Here common ratio $= \dfrac{2r}{r+3}$

   For the series to have a definite sum $\left|\dfrac{2r}{r+3}\right| < 1$

   $$\Rightarrow -1 < \frac{2r}{r+3} < 1 \Rightarrow 1 < r < 3$$

9. $a - b, b - c, c - a$ are in G.P.

   $$\Rightarrow \quad \frac{b-c}{a-b} = \frac{c-a}{b-c} \Rightarrow (b-c)^2 = (a-b).(c-a)$$

   $$\Rightarrow \quad a^2 + b^2 + c^2 = ab + bc + ca$$

   To show : $(a + b + c)^2 = 3(ab + bc + ca)$

   $$(a+b+c)^2 = a^2 + b^2 + c^2 + 2(ab+bc+ca)$$

   $$= 3(ab + bc + ca)$$